T0338442

NEURAL MODELS AND ALGORITHMS FOR DIGITAL TESTING

THE KLUWER INTERNATIONAL SERIES IN ENGINEERING AND COMPUTER SCIENCE

VLSI, COMPUTER ARCHITECTURE AND DIGITAL SIGNAL PROCESSING
Consulting Editor
Jonathan Allen

NEURAL MODELS AND ALGORITHMS FOR DIGITAL TESTING

by

Srimat T. Chakradhar
NEC Research Institute

Vishwani D. Agrawal
AT&T Bell Laboratories

Michael L. Bushnell
Rutgers University

Kluwer Academic Publishers
Boston/Dordrecht/London

Distributors for North America:
Kluwer Academic Publishers
101 Philip Drive
Assinippi Park
Norwell, Massachusetts 02061 USA

Distributors for all other countries:
Kluwer Academic Publishers Group
Distribution Centre
Post Office Box 322
3300 AH Dordrecht, THE NETHERLANDS

Library of Congress Cataloging-in-Publication Data

Chakradhar, Srimat T.
    Neural models and algorithms for digital testing./ by Srimat T. Chakradhar,
Vishwani D. Agrawal, Michael L. Bushnell.
        p.   cm. — (The Kluwer international series in engineering and computer
science ; SECS 140)
    Includes bibliographical references and index.
    ISBN 0-7923-9165-9 (acid-free paper)
    1. Logic circuits—Testing.   2. Automatic checkout equipment.   3. Digital
integrated circuits—Testing—Data processing.
    I. Agrawal, Vishwani D., 1943- .   II. Bushnell, Michael L. (Michael Lee), 1950- .
    III. Title.   IV. Series.
    TK7868.L6C44   1991
    621.39'5—dc20   91-15225
                                                                                    CIP

Printed on acid-free paper.

Printed in the United States of America

To

Rama and Srimat Tirumala Venkata Govindarajacharyulu

Premlata and Basudeo Sahai Agrawal

and

the memory of Bill Bushnell

Contents

Preface

This century has brought tremendous advances in computing. We have seen significant work on computer algorithms, programming languages, compilers and computing hardware. Have we reached the ultimate limit? Perhaps, unless we view the field of computing from a different angle and ask: *Can the machines designed to perform individually attain group synergy?* If not, we must find new computing disciplines for effective multiprocessing.

Neural networks, with their evolution-like computing capability, show hope for massive parallelism. Several problems are to be solved, however. Beside the hardware and software aspects, the problems also have to be coded as optimization or learning problems.

In this book, we present a novel solution of a difficult problem, namely, test generation for digital logic circuits. An optimization approach to this problem has only recently been attempted. We propose a new and unconventional modeling technique for digital circuits. The input and output signal states of a logic gate are related through an energy function such that the minimum energy states correspond to the gate's logic function. There are at least two advantages of this new approach. First, since the function of the circuit is expressed as a mathematical expression, new techniques can be used to solve a problem like test generation. Second, the non-causal form of the model allows effective use of parallel processing. We present the mathematical basis for our models and discuss their fundamental properties. Based on the same circuit models, we present test generation algorithms that can exploit fine-grain parallel computing, relaxation techniques, quadratic 0-1 programming and graph-theoretic techniques. We hope this unconventional solution will serve as a model for solving other problems.

In addition to its practical value, the proposed problem formulation leads to interesting theoretical contributions. As a further application of the model, we consider the intractability of the test generation problem. Using the neural network models, we present a new class of circuits in which this problem is solvable in polynomial time. This contribution is especially important since it leads to design styles for easily-testable digital circuits and provides a possible step toward design for testability. Furthermore, we discuss the application of the neural network models to other NP-complete problems. In particular, we present a new class of linear time solvable quadratic 0-1 programming cases. This result is significant because quadratic 0-1 programming is useful in solving practical problems such as Boolean satisfiability, traveling salesperson, VLSI layout and others. As an example, we show the application of the result in solving the independent set problem.

The present monograph describes the results of a research collaboration between the authors that was made possible by the research center for Computer Aids for Industrial Productivity (CAIP) at Rutgers University and by AT&T Bell Laboratories. A course offered by Mark Jones at the Rutgers University in 1987, that was attended by one of the authors (Chakradhar), may have started this work. We thank Prathima Agrawal and Apostolos Gerasoulis for several useful discussions and comments. We acknowledge Rutgers University and NEC Research Institute for providing excellent computing facilities. We also thank the Semiconductor Research Corporation and the National Science Foundation for their support of this work.

Finally, we thank our families for their excellent support during the course of the project.

Srimat Chakradhar
Vishwani Agrawal
Michael Bushnell

NEURAL MODELS AND
ALGORITHMS FOR
DIGITAL TESTING

Chapter 1

INTRODUCTION

"... I think it is reasonable to expect future neural networks to take on many of the aspects of an interconnected network of conventional computers. To me, that suggests a possible avenue toward the synthesis between computation and biological brain function that modern technology has long promised but has yet to achieve."

– A. Penzias in *Ideas and Information* (1989)

The origin of computing devices lies in computing problems. However, once a computing device is built, it requires algorithms through which the problems would be solved. Since the advent of electronic computers, the past decades have seen intense activity in the development of algorithms. Through this activity have emerged a class of problems that seem to require the cooperative use of multiple computers. Many problems in the design of large digital circuits fall in this category. Our emphasis, in this book, is on one such problem – the digital testing problem.

The *design* process transforms abstract functional specifications into a manufacturable assembly (or synthesis) of known parts. It also produces necessary data for driving the equipment that produces the assembly. *Verification* checks the correctness of design by analyzing its simulated performance. Due to errors in the fabrication process, every piece that is produced may not realize the intended function. *Testing* ensures the correctness of the manufactured product. Therefore, to obtain a functionally-correct product, both design and test data are necessary.

Work on parallelization of test generation algorithms has led to the belief that large-scale parallelism may not be advantageous. Obviously, the original developers of these algorithms that we are trying to parallelize had single-processor computers on their minds. We must recognize the important fact that an algorithm must provide the mapping between the problem and the computing machine architecture [7]. Our aim is to develop algorithms suitable for neural networks that are capable of massively parallel computing.

1.1   What Is Test Generation?

Rapid advances in integrated circuit technology have made it possible to fabricate digital circuits with a very large number of devices on a single chip. *Very large scale integration* (VLSI) is the fabrication of millions of components and interconnections at once by a common set of manufacturing steps. The testing process detects the physical defects produced during the fabrication of a VLSI chip. Such a chip is tested by a sequence of input stimuli, known as *test vectors*, that check for possible defects in the chip by producing observable faulty responses at primary outputs. *Test generation* involves the generation of test vectors to detect failures in the VLSI chip [1]. *Automatic test equipment* applies the test vectors to the input pins of the VLSI chip and compares the output pin responses with the expected responses. A normal requirement of these tests is that they detect a very high fraction of the modeled faults. The detected fraction of faults is called the *fault coverage*. Also, since the same set of test vectors will be applied to a large number (a million or more in many cases) of copies of the chip, short test sequences are desirable.

1.2   Why Worry About Test Generation?

The frequently-quoted advantages of VLSI are reduced system cost, improved performance, and greater reliability. These advantages, however, will be lost unless VLSI chips can be economically tested. An obvious reason for testing is to separate the good product from the faulty one. The dramatic increase in the ratio of the number of internal devices to input-output terminal pins of VLSI chips drastically reduces the *controllability* and *observability* of the circuit. Controllability refers to the ease of producing a specific internal signal value by applying signals to the circuit input terminals. Observability refers to the ease with which the state of internal signals can be deduced from the signals at the circuit output terminals.

A digital system consists of many printed wiring boards, each containing several VLSI chips. Testing such systems can be an overwhelming task and faster and more efficient test generation algorithms are needed. The variety of subtle failures that can occur in a VLSI chip further complicates the testing process. Complexity of test generation is enormous.

1.3 How About Parallel Processing?

No effective parallel architecture or algorithm for test generation has yet been found. At least two possibilities exist, however. First, a conventional uniprocessor algorithm can be parallelized by finding portions where it can be pipelined or directly executed in parallel. Second, the testing problem can be reformulated and new parallel methods can be developed. This book investigates the second approach and develops radically new circuit models and algorithms for test generation that can exploit massively parallel computers.

In the area of combinatorial optimization, good parallel methods have been designed to solve several standard types of problems [5]. Interestingly, most of this work has concentrated on speeding up the solutions of problems that could already be solved quite efficiently (i.e., in polynomial time) on a single-processor computer. The practical need for this speed up remains to be demonstrated. Certainly, the need to improve our ability to solve the hard (i.e., NP-complete) combinatorial problems is at least as pressing. Here, the parallelism offers a natural opportunity that has hardly been investigated. Even though one may never be able to change the exponential time to polynomial time, there is clearly a tremendous scope for parallelism as it may be the only way to obtain a solution in practice.

1.4 Neural Computing

We propose a new class of test generation algorithms that can exploit fine-grain parallel computing and relaxation techniques. Our approach radically differs from the conventional methods that generate tests for circuits from their gate level description. The circuit is modeled as a network of idealized computing elements, known as *neurons*, connected through bidirectional links. Our neuron is a binary (0-1) element. The relationship between the input and output signal states of a logic gate is expressed by an *energy function* such that the zero energy (also the minimum energy) states correspond to the gate's logic function. There are several advantages of this new approach. First, since the function of the circuit has been given a mathematical form, several new techniques can be used to solve problems like test generation. Second, the non-causal form of the model allows the use of parallel processing for compute-intensive design automation tasks. Using the neural network models, we formulate the test generation problem as an optimization problem. This is the first time such a formulation has been given for the test generation problem (Chapters 5 and 6).

Massive parallelism inherent in neural networks is exploited to generate tests. The digital circuit is represented as a bidirectional network of neurons. The circuit function is coded in the firing thresholds of neurons and the weights of interconnection links. This neural network is suitably reconfigured for solving the test generation problem. A fault is injected into the neural network and an energy function is constructed with global minima at test vectors.

We have simulated the neural network on a serial computer and determined the global minima of the energy function using a directed search technique augmented by probabilistic relaxation (Chapter 7).

Global minima of the energy function can also be determined by using analog neural networks. Since large-scale neural networks are not yet a reality, we simulated an analog neural network on a commercial neurocomputer to generate tests (Chapter 8). Our neurocomputer contains special-purpose hardware for high-speed simulation of neural networks. Preliminary results on combinational circuits confirm the feasibility of these techniques.

1.5 A Novel Solution

A novel test generation method using *quadratic 0-1 programming* can be devised. The energy function, a quadratic function of 0-1 variables, is split into two sub-functions, a homogeneous posiform and an inhomogeneous posiform. A posiform is a pseudo-Boolean quadratic function in which all terms have positive coefficients (for precise definitions, see Section 9.2). The minimum of the energy function is the sum of the minima of the two sub-functions, each having the minimum value of 0. We obtain a minimizing point of the homogeneous posiform, in time complexity that is linear in the number of sub-function terms, and check if the other sub-function also becomes 0. When both sub-functions simultaneously become 0, we have a test vector. Note that this is an entirely *new* method for minimizing quadratic 0-1 functions. We discuss several speedup techniques using transitive closure and other graph properties. This approach is also effective in identifying redundant faults in the circuit. Such faults do not affect the circuit behavior and, hence, cannot be detected by any test. However, which faults are redundant is not known *a priori* and test generators spend considerable amounts of resources to identify them. Preliminary results on combinational circuits confirm the feasibility of the proposed test generation technique (Chapters 9 and 10).

1.6 Polynomial Time Test Problems

Test generation for general logic circuits is an extremely difficult problem. A practical approach is to identify special circuits that are easily-testable. This leads to the evolution of design styles for synthesizing easily-testable circuits. Very little prior work has been done on identifying such circuits. We present a new family of circuits, called the (k, K)-*circuits*, in which the test generation problem is solvable in polynomial time. Unlike prior easily-testable classes of circuits [3], the (k, K)-circuits can have some reconvergent fanout. This contribution is important for yet another reason. In all currently known test generation algorithms, it has not been possible to precisely characterize the class of circuits they work best for. Such information is useful for two reasons: (1) we could select a particular test generation algorithm for a given circuit and, (2) we could tailor our synthesis procedures to design easily-testable circuits. In Chapter 11, we present a test generation algorithm and show that it works best on the (k, K)-circuits.

1.7 Application to Other NP-complete Problems

Problem solving using neural networks requires formulation of the problem as an optimization or a learning problem. The neural models are useful in reformulating problems as optimization problems that can be solved using neural networks. As an illustration, we show the use of neural models in formulating the test generation problem as an optimization problem in Chapter 6. Similar techniques can be used to solve the Boolean satisfiability problem, the maximum independent set problem and others using neural networks.

A novel application of the proposed modeling technique is in the identification of new, easily-solvable instances of NP-complete problems. We consider quadratic 0-1 programming, a representative NP-complete problem, and identify its linear time solvable instances. The significance of this result stems from the fact that the VLSI layout problem, the Boolean satisfiability problem, the traveling salesperson problem and many other combinatorial optimization problems can be solved via quadratic 0-1 programming [2, 6]. Our research has resulted in a new linear time algorithm to find the minimum of $f(x) = x^T Q x + c^T x$ with $x \in \{0, 1\}^n$, when a certain graph defined by f is transformable to a combinational circuit of inverters and 2-input AND, OR, NAND and NOR logic gates. Here, Q is an $n \times n$ symmetric matrix of constants, c is a vector of constants, x is a vector of n Boolean variables, and x^T is the transpose of x. The graph of f, $G_f(\mathcal{V}, \mathcal{E})$ where $\mathcal{V}$ is the vertex set and $\mathcal{E}$ the edge set of the graph, is defined as follows:

$$\mathcal{V} = \{x_1, x_2, \ldots, x_n\}$$
$$(x_i, x_j) \in \mathcal{E} \Leftrightarrow Q_{ij} \neq 0 \; (i \in \mathcal{V}, j \in \mathcal{V})$$

Whenever a combinational circuit is realizable, the solution is found by assigning arbitrary 0-1 values to the primary inputs and performing logic simulation. Another contribution is related to the complexity of restricted versions of quadratic 0-1 programming. Finding the minimum of f is an NP-complete problem [4]. It is, however, solvable in polynomial time when all elements of Q are restricted to be non-positive [8]. We have shown that the problem remains NP-complete even for the special case when all elements of Q are positive (Chapter 12).

Another application of the neural models is in solving problems using logic hardware. A possible hardware solution of the quadratic 0-1 programming problem is illustrated in Chapter 12. In Chapter 13, we sketch out a possible application of neural models to solve a difficult graph problem, namely, the independent set problem. This chapter reports on our ongoing work.

1.8   Organization of the Book

The contribution of this research spans three areas, namely, test generation, massively parallel computing using neural networks and mathematical programming. Since an understanding of this work will require basic knowledge of all the three areas, we devote Chapters 2, 3 and 4 to such background. A knowledgeable reader is advised to skip the unnecessary detail. Preliminary material on mathematical programming is reviewed in Section 9.2.

In Chapter 5, we present our novel model of logic circuits using neural networks. We present the mathematical basis for the model and discuss its fundamental properties. In Chapter 6, we formulate test generation as an energy minimization problem. Using the neural models, two new techniques are developed for test generation. Chapters 7 and 8 illustrate the use of relaxation techniques and fine grain parallel computing for test generation. Chapter 9 presents a new, discrete non-linear programming technique for test generation. Preliminary results confirm the feasibility of both techniques. Chapter 9 also introduces the application of graph-theoretic techniques to test generation and Chapter 10 provides a detailed exposition on the application of transitive closure to digital testing.

Since the general problem of test generation is intractable, we investigate the existence of classes of circuits for which this problem can be solved in polynomial time. This theoretical work, we believe, will lead to the evolution of new design styles for easily-testable circuits. Chapter 11 uses the neural network models and presents a new class of circuits for which the test generation problem is solvable in polynomial time.

In Chapter 12, we discuss an application of the neural network model to other NP-complete problems. In particular, we present a new class of quadratic 0-1 programming problems that are solvable in linear time. Applications here are not restricted to test generation or other *Computer-Aided Design* (CAD) problems. In Chapter 13, we explore the possibility of using the neural models for solving difficult graph problems. In Chapter 14, we suggest possible future applications of the the neural network model and outline open problems.

References

[1] V. D. Agrawal and S. C. Seth. *Test Generation for VLSI Chips.* IEEE Computer Society Press, Los Alamitos, CA, 1988.

[2] M. W. Carter. The Indefinite Zero One Quadratic Problem. *Discrete Applied Mathematics,* 7(1):23–44, January 1984.

[3] H. Fujiwara. Computational Complexity of Controllability/Observability Problems for Combinational Circuits. *IEEE Transactions on Computers,* C-39(6):762–767, June 1990.

[4] M. R. Garey and D. S. Johnson. *Computers and Intractability: A Guide to the Theory of NP-Completeness.* W.H. Freeman & Company, San Francisco, 1979.

[5] A. Gibbons and W. Rytter. *Efficient Parallel Algorithms.* Cambridge University Press, U.K., 1988.

[6] P. L. Hammer, P. Hansen, and B. Simeone. Roof Duality, Complementation and Persistence in Quadratic 0-1 Optimization. *Mathematical Programming,* 28(2):121–155, February 1984.

[7] E. V. Krishnamurthy. *Parallel Processing Principles and Practices.* Addison-Wesley, Sydney, 1989.

[8] J. C. Picard and H. D. Ratliff. Minimum Cuts and Related Problems. *Networks,* 5(4):357–370, October 1975.

Chapter 2

LOGIC CIRCUITS AND TESTING

"Testing of large circuits can be very testing."

This chapter is a tutorial on logic circuit modeling, fault modeling and test generation [5]. Concepts of testing, as used in later chapters, are briefly reviewed. Readers having a basic understanding of these concepts may choose to skip this chapter.

2.1 Logic Circuit Preliminaries

Signals and Gates: A *signal* is a Boolean variable that may assume only one of the two possible values represented by the symbols 0 and 1. A *gate* is simply an electronic circuit which operates on one or more input signals to produce an output signal. Typical gates used in logic circuits perform Boolean functions like *AND, OR, NAND, NOR* and *NOT.* An AND gate performs the conjunction ($\wedge$) of the input signals. For example, if x_1 and x_2 are the input signals of the AND gate, the output signal x_3 is the conjunction of its two input signals, i.e., $x_3 = x_1 \wedge x_2$. Similarly, an OR gate performs the disjunction ($\vee$) of the input signals. The complement of a Boolean variable is physically realized by a NOT gate, also called an inverter. For example, if x_1 is the input signal to an inverter, the output signal $x_2 = \overline{x}_1$. A NAND gate performs the complement of the conjunction of its input signals. For example, if x_1 and x_2 are the input signals to the NAND gate, then the output signal $x_3 = \overline{x_1 \wedge x_2}$. A NOR gate performs the complement of the disjunction of the input signals. Two other types of gates, *XOR* and *XNOR,* are often used in logic circuits. If x_1 and x_2 are the input signals to the XOR (exclusive-OR) gate, the output signal x_3 is given by $x_3 = x_1 \oplus x_2 = (x_1 \wedge \overline{x}_2) \vee (\overline{x}_1 \wedge x_2)$. The XNOR (exclusive-NOR) gate is an inversion of XOR and produces $x_3 = \overline{x_1 \oplus x_2}$.

Logic Circuits: Logic circuits are modeled as interconnections of gates. For simplicity of presentation, we will only consider gates with one or two input signals and one output signal. We will refer to a particular gate in the circuit by its output signal name. For example, Figure 2.1 shows a logic circuit with three gates. The AND gate x_3 has x_1 and x_2 as its input signals.

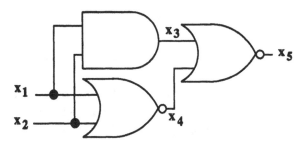

FIGURE 2.1: A combinational circuit example.

The NOR gate x_4 also has x_1 and x_2 as its input signals and NOR gate x_5 has x_3 and x_4 as input signals. Signals that are not the output signals of any gate in the circuit are called *primary inputs*. For example, signals x_1 and x_2 are the primary inputs of the circuit in Figure 2.1. Signals that realize the output functions of the circuit are called *primary outputs*. For example, signal x_5 is the primary output of the circuit in Figure 2.1. In general, a circuit can have several primary input and primary output signals.

Logic circuits can be categorized as *combinational* or *sequential*. Consider a logic circuit C consisting of signals $x_1, \ldots, x_n$. Let $H_C(\mathcal{V}_C, \mathcal{E}_C)$, where $\mathcal{V}_C$ is the vertex set and $\mathcal{E}_C$ the edge set, be the *circuit graph* associated with C. $\mathcal{V}_C = \{x_1, x_2, \ldots, x_n\}$ and there is an arc $(x_i, x_j) \in \mathcal{E}_C$ ($1 \leq i \leq n$ and $1 \leq j \leq n$) if there is a gate x_j with x_i as an input signal. If H_C has no directed *cycles* [7], C is a combinational circuit. Otherwise, C is a sequential circuit. For example, it can easily be verified that the circuit graph corresponding to the circuit in Figure 2.1 has no directed cycles and, therefore, it is a combinational circuit. In the sequel, we will only deal with combinational circuits.

2.2 Test Generation Problem

An input *vector* is a set of values for all primary input signals. A *fault* is a fabrication-related failure and the generation of a test for a given fault involves searching among the set of possible input vectors of the circuit until a *test vector* is found that differentiates the fault-free circuit from the faulty one. Primary inputs are the only points where we can apply test signals to the circuit and primary outputs are the only points where we must observe the effect of the fault. Thus, a test for a given fault causes the logic value on at least one primary output of the faulty circuit to differ from its expected value in the fault-free circuit. Test generation is the task of finding a set of test vectors that will fully test the circuit for a given set of faults. *Automatic test generation* (ATG) algorithms can be programmed on digital computers for generating test vectors.

2.2.1 Fault Modeling

Testing that uses all possible input vectors will, of course, reveal any faulty circuit. However, such a procedure will be too expensive for large circuits with many primary inputs (a circuit with n primary input signals will have 2^n possible input vectors). To simplify the testing of large circuits we make assumptions about the possible failures. The large number and complex nature of physical failures dictate that a practical approach to testing should avoid working directly with physical failures. In most cases, one is not concerned with discovering the exact physical failure; what is desired is merely to determine the existence (or absence) of any physical failure. One approach is to describe the effects of physical failures at some higher level (logic, register transfer, functional block, etc.). This abstraction is called *fault modeling*.

In the classical gate-level *stuck-at* fault model, the effects of physical failures are assumed to cause the inputs and outputs of logic gates to be permanently "stuck" at logic 0 or 1 (commonly called *s-a-0* or *s-a-1* faults). Much of the work in testing has used this model. It is easy to see that a circuit with n signals can have as many as $3^n - 1$ possible faults since each signal can be s-a-0, s-a-1, or fault-free. This is a large number. For simplicity, it is assumed that only one stuck-at fault will be present at one time. With this single fault assumption, a circuit with n signals can have at most $2n$ faults. Further reduction in the number of faults occurs through *fault collapsing* when *equivalent classes* of faults are recognized. Two faults are equivalent if they are detected by exactly the same test vectors. The single-stuck-at fault model might seem artificial on the surface, but it is quite useful. Any set of test vectors that we use for detecting a large percentage of single stuck-at faults, in practice, also detects a high percentage of all failures [6].

2.2.2 Problem Definition

The problem of test generation for a stuck-at fault in a combinational circuit with n primary input signals can be formulated as a search in the n-dimensional 0-1 state space of primary input vectors. Consider the combinational circuit of Figure 2.2. The primary input signals are labeled $x_1, \ldots, x_n$. The primary output signals are labeled $f_1, \ldots, f_m$. Recall that the primary input and output signals are Boolean variables. Suppose that

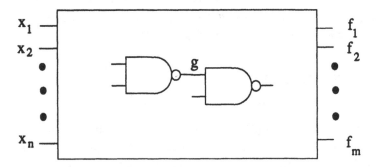

FIGURE 2.2: A generic combinational circuit.

the objective is to generate a test for the stuck-fault s-a-0 on some signal g that is not a primary input. We will refer to the fault as g s-a-0 (read as signal g stuck-at-0). Signal g can be expressed as a Boolean function $g = G(x_1, x_2, \ldots, x_n)$ of the primary inputs. Similarly, primary output signal f_j can be expressed as a Boolean function of signal g and the primary input signals, as $f_j = F_j(g, x_1, x_2, \ldots, x_n)$, where $1 \leq j \leq m$. The problem of generating a test for g s-a-0 can be recast as the problem of finding a Boolean vector $(x_1, x_2, \ldots, x_n)$ that simultaneously satisfies two Boolean equations given below:

$$G(x_1, x_2, \ldots, x_n) = 1$$
$$F_j(1, x_1, x_2, \ldots, x_n) \oplus F_j(0, x_1, x_2, \ldots, x_n) = 1$$

for at least one j in the range $1 \leq j \leq m$. The operator $\oplus$ denotes the exclusive-or operation. The two equations for the fault g s-a-1 are similar except that the first equation becomes $G(x_1, \ldots, x_n) = 0$. This formulation is also referred to as the *Boolean difference* [25].

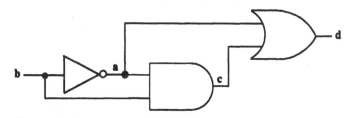

FIGURE 2.3: A circuit with a redundant fault c s-a-0.

Example 2.1: Suppose the objective is to generate a test for the fault c s-a-1 in Figure 2.3. We can restate the problem in the Boolean difference form as follows:

1. Express signal c as a Boolean function of the primary input signals of the circuit, i.e., $c = G(b) = b \wedge \bar{b}$.

2. Express the primary output signal d as a Boolean function of signal c and the primary inputs, i.e., $d = F(c, b) = c \vee \bar{b}$.

3. Find a logic value of signal b that satisfies the following Boolean equations:

$$G(b) = b \wedge \bar{b} = 0$$
$$F(0, b) \oplus F(1, b) = \bar{b} \oplus 1 = 1$$

Since b is a Boolean variable, there are only two possible values for b. It can easily be verified that $b = 1$ satisfies the above equations and, hence, it is a test for the fault.

Test generation is the task of generating a test set that reveals all detectable single stuck-at faults. We say detectable faults because a circuit can have undetectable faults, also called *redundant faults*. For example, suppose the objective is to generate a test for the fault c s-a-0 in Figure 2.3. Stating this as a search problem, we must find a logic value of signal b that satisfies the two Boolean equations:

$$G(b) = b \wedge \bar{b} = 1$$
$$F(0, b) \oplus F(1, b) = \bar{b} \oplus 1 = 1$$

Clearly, no value of signal b can satisfy these equations. Hence, there is no test for the fault c s-a-0. Redundant faults do not affect the normal operation of the circuit. Their identification, however, is at least as hard as generating a test for a detectable fault. The presence of a redundant fault implies that the Boolean function realized by the combinational circuit can also be realized by another combinational circuit having fewer gates.

2.2.3 Complexity of Test Generation

The number of primary inputs and the number of signals in the combinational circuit are generally considered as the the *input size* for the test generation problem. Ibarra and Sahni [18] show that the test generation problem for combinational circuits is NP-complete. This means that no test generation algorithm with a polynomial time complexity is likely to exist [15]. The non-polynomial (or exponential) time complexity here refers to the *worst-case* effort of test generation for any fault in the circuit. In fact, very restricted versions of the problem are NP-complete [11, 12, 14].

2.3 Test Generation Techniques

There has been a heavy emphasis on automatic test pattern generation for combinational circuits since these techniques can also be used to test specially designed sequential circuits [1]. A wide range of techniques have been proposed for combinational circuit test generation. At one end of the spectrum are the exhaustive and random techniques. If the number of primary inputs of the circuit is small, exhaustive test generation that applies all possible input vectors may be an obvious candidate. Random test generation [3] is another simple technique that probabilistically generates input vectors and verifies if they detect any faults in the circuit. If a random vector detects any fault, then it is retained as a test. The number of random vectors needed for high fault coverage can be very large in some cases.

At the other end of the spectrum are the deterministic techniques that can be partitioned into two groups. One group consists of the *algebraic* algorithms that use the the Boolean difference formulation (Section 2.2.2) to symbolically solve the problem [11, 25]. These algorithms have not been very practical for two reasons: (1) Symbolic solutions require excessive storage and (2) Heuristics required to symbolically solve practical test generation problems are not available. Yet this group of algorithms serves the important purpose of illuminating the fundamental nature of the testing problem [2].

The second group of *structural* algorithms solve the problem by exploiting the transistor or gate-level representation of the circuit. These algorithms systematically enumerate the search space using the branch-and-bound method and employ heuristics to guide the search process. The first structural algorithm to be proposed was the D-algorithm [23]. Here, every signal in the circuit is considered as a decision point in the search process. If a decision results in some signal being assigned both logic values (i.e., 0 and 1), we will have to retract that decision and make an alternative choice. This process is called *backtracking*. The D-algorithm performs particularly poorly on circuits that require an excessive amount of backtracking. The *Path Oriented Decision Making* (PODEM) algorithm [17] offers a significant improvement. Here, only primary input signals are considered as decision points in the search process. Heuristics are used to dynamically determine the next decision point. PODEM implementations are known to run an order of magnitude faster than the D-algorithm on most circuits. FAN [10], TOPS [19] and SOCRATES [24] incorporate heuristics to accelerate the basic PODEM algorithm. They perform extensive topological analyses of circuit connectivity to develop better heuristics for the branch-and-bound process. In another recent technique, to accelerate test generation, search-states are saved at every step. Equivalent search-states are identified and they are used to prune the search space [16].

The CONT [28] algorithm pursues a different approach. If the target fault is not detectable by the current input vector, CONT switches the target to another undetected fault. Candidate faults for target switching are identified by interleaving fault simulation steps with incremental test generation steps. In the CONT-2 algorithm [29], all unspecified inputs are randomly assigned prior to fault simulation.

Approximation algorithms may not guarantee a test but they have been proven to be effective for many practical circuits. Simulation-based directed search is one such technique that makes use of logic and fault simulation, and is particularly effective for sequential circuit test generation [8]. The CONTEST algorithm [4] uses concurrent fault simulation and the TV-SET algorithm [9] is based upon a new threshold-value model of the circuit. Both methods include gate delays to correctly determine tests for sequential circuits. Expert systems have also been used for test generation but with limited success [26].

2.4 Parallelization

One common theme that all the above techniques share is that they are developed for single-processor computers that execute the algorithm as a single stream of steps. Kramer [20] developed a test generator using the Connection Machine. However, due to the exhaustive search used, only circuits with 15 to 18 input lines could be processed.

In most of the recent work, three approaches are evident:

- Fault parallelism.

- Search parallelism.

- Element-level parallelism.

In the case of fault parallelism, the fault list is partitioned for distribution among processors. To avoid interprocessor communication, the processors may independently generate tests. However, duplication of work occurs if the test generated in one processor can detect a fault that is assigned to another processor. A close to ideal speedup is only possible if the fault list is carefully partitioned and interprocessor communication delay can be neglected [13].

In search parallelism, all processors in the multiprocessing system work in unison to find a test for the same given fault [22]. Every processor has a copy of the circuit. The amount of memory available on a single processor in the multiprocessor system limits the size of circuits that can be tested using this method. The search space for the problem is divided into disjoint sub-spaces and each processor searches one sub-space for a test vector. If a processor finds a test vector, it sends messages to all other processors to abort further search. This is an obvious way to parallelize any branch-and-bound method but some ingenuity is required to isolate sub-spaces that have a higher likelihood of containing a solution. By searching such sub-spaces, we have a better chance of quickly converging on to a test vector. Another implementation of search space partitioning has recently been reported on an eight-processor Sequent computer [27]. Unlike other work, here the Boolean satisfiability method is used for test generation [21].

In the element-level parallelism, the gates of the circuit are distributed on multiple processors. A recent study [30] concludes that fault parallelism should be preferred over the element-level parallelism. The main reason is the excessive amount of interprocessor communication required in the element-level parallelism.

References

[1] M. Abramovici, M. A. Breuer, and A. D. Friedman. *Digital Systems Testing and Testable Design*. Computer Science Press, New York, NY, 1990.

[2] V. K. Agarwal and A. S. Fung. Multiple Fault Testing of Large Circuits by Single Fault Test Sets. *IEEE Transactions on Computers*, C-30(11):855–865, November 1981.

[3] V. D. Agrawal. When to Use Random Testing. *IEEE Transactions on Computer-Aided Design*, C-27(11):1054–1055, November 1978.

[4] V. D. Agrawal, K. T. Cheng, and P. Agrawal. A Directed-Search Method for Test Generation using a Concurrent Simulator. *IEEE Transactions on Computer-Aided Design*, 8(2):131–138, February 1989.

[5] V. D. Agrawal and S. C. Seth. *Test Generation for VLSI Chips*. IEEE Computer Society Press, Los Alamitos, CA, 1988.

[6] V. D. Agrawal, S. C. Seth, and P. Agrawal. Fault Coverage Requirement in Production Testing of LSI Circuits. *IEEE Journal of Solid-State Circuits*, SC-17(1):57–61, February 1982.

[7] A. V. Aho, J. E. Hopcroft, and J. D. Ullman. *The Design and Analysis of Computer Algorithms*. Addison-Wesley Publishing Company, Reading, MA, 1974.

[8] K. T. Cheng and V. D. Agrawal. *Unified Methods for VLSI Simulation and Test Generation*. Kluwer Academic Publishers, Boston, 1989.

[9] K. T. Cheng, V. D. Agrawal, and E. S. Kuh. A Simulation-Based Method for Generating Tests for Sequential Circuits. *IEEE Transactions on Computers*, C-39(12):1456–1463, December 1990.

[10] H. Fujiwara. FAN: A Fanout-Oriented Test Pattern Generation Algorithm. In *Proceedings of the IEEE International Symposium on Circuits and Systems*, pages 671–674, July 1985.

[11] H. Fujiwara. *Logic Testing and Design for Testability*. MIT Press, Cambridge, Massachusetts, 1985.

[12] H. Fujiwara. Computational Complexity of Controllability/Observability Problems for Combinational Circuits. *IEEE Transactions on Computers*, C-39(6):762–767, June 1990.

[13] H. Fujiwara and T. Inoue. Optimal Granularity of Test Generation in a Distributed system. *IEEE Transactions on Computer-Aided Design*, 9(8):885–892, August 1990.

[14] H. Fujiwara and S. Toida. The Complexity of Fault Detection Problem for Combinational Circuits. *IEEE Transactions on Computers*, C-31(6):555–560, June 1982.

[15] M. R. Garey and D. S. Johnson. *Computers and Intractability: A Guide to the Theory of NP-Completeness*. W.H. Freeman & Company, San Francisco, 1979.

[16] J. Giraldi and M. L. Bushnell. EST: The New Frontier in Automatic Test-Pattern Generation. In *Proceedings of the 27th ACM/IEEE Design Automation Conference*, pages 667–672, June 1990.

[17] P. Goel. An Implicit Enumeration Algorithm to Generate Tests for Combinational Logic Circuits. *IEEE Transactions on Computers*, C-30(3):215–222, March 1981.

[18] O. H. Ibarra and S. K. Sahni. Polynomially Complete Fault Detection Problems. *IEEE Transactions on Computers*, C-24(3):242–249, March 1975.

[19] T. Kirkland and M. R. Mercer. A Topological Search Algorithm For ATPG. In *Proceedings of the 24th ACM/IEEE Design Automation Conference*, pages 502–508, June 1987.

[20] G. Kramer. Employing Massive Parallelism in Digital ATPG Algorithms. In *Proceedings of the IEEE International Test Conference*, pages 108–121, 1983.

[21] T. Larrabee. Efficient Generation of Test Patterns Using Boolean Difference. In *Proceedings of the IEEE International Test Conference*, pages 795–801, August 1989.

[22] S. Patil and P. Banerjee. A Parallel Branch and Bound Algorithm for Test Generation. *IEEE Transactions on Computer-Aided Design*, 9(3):313–322, March 1990.

[23] J. P. Roth, W. G. Bouricius, and P. R. Schneider. Programmed Algorithms to Compute Tests to Detect and Distinguish Between Failures in Logic Circuits. *IEEE Transactions on Electronic Computers*, EC-16(5):567–580, October 1967.

[24] M. H. Schulz, E. Trischler, and T. M. Sarfert. SOCRATES: A Highly Efficient Automatic Test Pattern Generation System. *IEEE Transactions on Computer-Aided Design*, 7(1):126–136, January 1988.

[25] E. F. Sellers, M. Y. Hsiao, and L. W. Bearnson. Analyzing Errors with the Boolean Difference. *IEEE Transactions on Computers*, C-17(7):676–683, 1968.

[26] N. Singh. *An Artificial Intelligence Approach to Test Generation*. Kluwer Academic Publishers, Boston, 1987.

[27] V. Sivaramakrishnan, S. C. Seth, and P. Agrawal. Parallel Test Pattern Generation using Boolean Satisfiability. In *Proceedings of the 4th CSI/IEEE International Symposium on VLSI Design, New Delhi*, pages 69–74, January 1991.

[28] Y. Takamatsu and K. Kinoshita. CONT: A Concurrent Test Generation System. *IEEE Transactions on Computer-Aided Design*, 8(9):966–972, September 1989.

[29] Y. Takamatsu and K. Kinoshita. Extended Selection of Switching Target Faults in CONT Algorithm for Test Generation. *J. Electronic Testing: Theory and Applications*, 1(3):183–189, October 1990.

[30] N. A. Zaidi and S. A. Szygenda. Design and Analysis of a Hardware Automatic Test Generation System. In A. P. Ambler, P. Agrawal, and W. R. Moore, editors, *CAD Accelerators*, pages 103–114. North-Holland, Amsterdam, 1991.

Chapter 3

PARALLEL PROCESSING PRELIMINARIES

*"If we shred a page into a thousand pieces, we can print
one letter on each piece in parallel. But then, putting the
printed page together will take a long time."*

Parallel processing is viewed as a way to overcome the limitations of single-processor systems. Two distinct styles of parallel computing are emerging. One method uses an array of cooperating processors, suitably synchronized by a clocking mechanism, such that each develops a part of the solution that it shares with other neighbors. The other method uses *neural networks* and is a radically different way of parallel processing. These networks are interconnections of *analog* computing elements that cooperatively evolve toward a configuration, interpreted as a solution to the problem. There is no clock and the time evolution of the network follows a trajectory described by a set of differential equations. Such computation, that will perhaps mimic the behavior of the human brain, may be the way of the future [13].

In this chapter, we briefly review the basic parallel processing concepts [11] and in the next chapter we provide an introduction to neural networks. Readers with a basic understanding of these topics can skip these chapters.

3.1 Synchronous Parallel Computing

In this model of parallel processing, a programmed algorithm is regarded as a collection of sequential units of computation (*tasks*) that are constrained in their execution by dependencies, i.e., certain tasks must be completed before some other tasks can begin. A dependence exists between two tasks either because they share data (*data dependence*) or because one task determines whether or not the other task is executed (*control dependence*). There are three fundamental steps to executing a program on a multiprocessor system: identifying the parallelism in the program, *partitioning* the program into tasks and *scheduling* the tasks on processors [9].

3.1.1 Program Representation

A program can be represented by a directed graph $G(V, E)$, called the *dependence graph* [7], with vertex set V and edge set E, where $V = \{T_1, T_2, ..., T_n\}$ is a set of tasks and there is a directed edge from T_i to T_j if task T_i must be completed before task T_j can begin. With each vertex (task) in the graph, we associate a real number that represents the execution time of the task. An important issue in the construction of the graph for a given program is the determination of execution times for various tasks. For many programs, it is not possible to determine the exact task execution time since it may depend on the input data. Execution profile information averaged over different data inputs is frequently used to estimate the execution times of the tasks as discussed by Sarkar [12].

3.1.2 Identification of Program Parallelism

In general, the amount of parallelism in a program depends upon the algorithm designed by the programmer. However, the identification of parallelism in the program may also depend upon the programming language and the compiler. Partitioning and scheduling problems are intimately related to the architecture of the target multiprocessor.

3.1.3 Partitioning and Scheduling

It is important to determine the absolute or relative execution times of tasks in the program dependence graph, either by execution profiling or some other means. The partitioning problem then reduces to determining a partition of the program into tasks so as to minimize the parallel execution time of the program for a given target multiprocessor. Partitioning is necessary to ensure that the granularity of the parallel program is coarse enough for the target multiprocessor, without losing too much parallelism.

The scheduling problem is to assign tasks in the partitioned program to processors to minimize the overall parallel execution time. Scheduling is necessary for achieving an efficient processor utilization and for optimizing inter-processor communication in the target multiprocessor system.

It is desirable that partitioning and scheduling be performed automatically, so that the same parallel program can execute efficiently on different multiprocessor systems. There are two distinct approaches to solving the partitioning and scheduling problems. The first approach partitions the program into tasks during compilation and the tasks are scheduled on processors at run time. In the second approach, the partitioning of the program and the scheduling of tasks on processors are both performed at compile-time. The task partitions are usually restricted so that the inter-task dependencies are *acyclic*. This greatly simplifies the scheduling problem.

3.1.4 Performance Measures

We define parallel processor system performance in terms of two quantities – *speedup* and *efficiency*. The speedup S_p for a given program is the ratio $\frac{T_1}{T_p}$, where T_1 is the execution time of the program on a single processor, and T_p is the execution time of the program on multiprocessor system with p processors. *Ideal* speedup is p or linear in the number of processors. The efficiency of the multiprocessor system when executing a particular program, denoted as E_p, is the ratio of the observed speedup to the number of processors: $\frac{S_p}{p}$. While these measures are frequently used, other performance metrics have also been proposed. For example, Karp and Flatt [10] have recently suggested the use of the experimentally measured *serial fraction*, a characteristic of the program, in performance evaluation.

3.1.5 Solving Problems Using Multiprocessors

To solve a problem using multiprocessors, we express the corresponding algorithm as a dependence graph. We then partition and schedule this graph on a target multiprocessor by using appropriate techniques. See, for example, Sarkar [12]. In spite of our best efforts to evenly distribute the work on processors, there can be a statistical variation of computational activity among them. Such variation, that can significantly reduce the gain of parallel processing, has been studied by Agrawal and Chakradhar [4, 5, 6]. They define *activity* as the probability that a computing statement requires processing due to a change in its input data. The actual speedup is found to reduce by a factor equal to the activity over the ideal speedup. They also study the cases of unequal task division and of massive parallelism. These results are in remarkable agreement with those obtained for parallel logic simulation [1].

3.2 Parallel Test Generation

Techniques for parallel discrete event simulation are described in a recent survey [8]. Multiprocessor systems, built for logic simulation [3] and fault simulation [2], are currently in use. Parallelization of test generation, however, has not reached the same level of maturity. Recent work on parallel computing for test generation is discussed in Section 2.4.

Many researchers believe that massive parallelism and new algorithms rather than the raw speed of individual processors may provide the computational power necessary to carry out the increasingly complex calculations in combinatorial optimization and artificial intelligence. We investigate the use of massive parallelism in neural networks for solving the test generation problem. Our approach will require a radically different representation of digital circuits and a new formulation of the test generation problem. This is because the causal form of the conventional circuit representation techniques limits the use of parallelism.

References

[1] P. Agrawal. Concurrency and Communication in Hardware Simulators. *IEEE Transactions on Computer-Aided Design*, CAD-5(4):617–623, October 1986.

[2] P. Agrawal, V. D. Agrawal, K. T. Cheng, and R. Tutundjian. Fault Simulation in Pipelined Multiprocessor Systems. In *Proceedings of the IEEE International Test Conference*, pages 727–734, August 1989.

[3] P. Agrawal and W.J. Dally. A Hardware Logic Simulation System. *IEEE Transactions on Computer-Aided Design*, CAD-9(1):19–29, January 1990.

[4] V. D. Agrawal and S. T. Chakradhar. Logic Simulation and Parallel Processing. In *Proceedings of the IEEE International Conference on Computer-Aided Design*, pages 496–499, November 1990.

[5] V. D. Agrawal and S. T. Chakradhar. Performance Estimation in a Massively Parallel System. In *Proceedings of the ACM Supercomputing Conference*, pages 306–313, November 1990.

[6] V. D. Agrawal and S. T. Chakradhar. Performance Evaluation of a Parallel Processing System. In *ISMM Proceedings of the International Conference on*

Parallel and Distributed Computing, and Systems, pages 212–216, October 1990.

[7] D. J. Kuck et. al. Dependence Graphs and Compiler Optimizations. In *Proceedings of the 8th ACM Symposium on Principles of Programming Languages*, pages 207–218, January 1981.

[8] R. M. Fujimoto. Parallel Discrete Event Simulation. *Communications of the ACM*, 33(10):31–53, October 1990.

[9] A. Gerasoulis and I. Nelken. Gaussian Elimination and Gauss Jordan on MIMD Architectures. Technical Report LCSR-TR-105, Department of Computer Science, Rutgers University, New Brunswick, New Jersey 08855, 1989.

[10] A. H. Karp and H. P. Flatt. Measuring Parallel Processor Performance. *Communications of the ACM*, 33:539–543, May 1990.

[11] E. V. Krishnamurthy. *Parallel Processing Principles and Practices*. Addison-Wesley, Sydney, 1989.

[12] V. Sarkar. *Partitioning and Scheduling Parallel Programs for Multiprocessors*. Research Monographs in Parallel and Distributed Computing, The MIT Press, Cambridge, Massachusetts, 1989.

[13] B. Soucek and M. Soucek. *Neural and Massively Parallel Computers – The Sixth Generation*. John Wiley & Sons, Wiley Interscience publications, New York, NY, 1988.

Chapter 4

INTRODUCTION TO NEURAL NETWORKS

"When you see a picture, signals from different parts of the picture are simultaneously picked up by thousands of optic nerves for processing in the brain all at once. Maybe, that is how a picture is worth a thousand words"

A *neuron* is a simple computing element and a *neural network* is an interconnection of such computing elements [2, 3]. In this chapter, we review an electrical model of the neuron, analyze the mechanism of computing in a network of such neurons and show that this computation is equivalent to the minimization of a certain cost function known as the *energy* of the network [1]. We also review a possible implementation of a neuron as an analog circuit and discuss general techniques to construct neural networks for combinatorial optimization problems.

4.1 Discrete Model of Neuron

A neuron is viewed as a computing element that can assume one of two possible states: 0 or 1. Figure 4.1 shows the locality of a neuron in a network. Associated with neuron i is a real number I_i, the *threshold* of the neuron, and an output value V_i. A neuron is connected to its neighbors through *links*. We associate a real number T_{ij} with the link from neuron

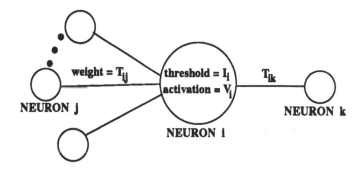

FIGURE 4.1: A typical neuron.

j to neuron i. We assume that $T_{ij} = T_{ji}$. Furthermore, $T_{ii} = 0$ since the neurons do not have self-feedback. A neuron receives inputs from its neighbors, computes an output as a function of the inputs and sends the output value to its neighbors.

A natural representation for a neural network is a weighted graph, with neurons represented as vertices, and connections as weighted edges. Each vertex is also assigned a threshold. As an example, consider the neural network shown in Figure 4.2. The network has four neurons labeled a, b, c

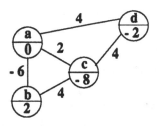

FIGURE 4.2: Example neural network.

and d, whose thresholds are 0, 2, -8 and -2, respectively. There are five weighted edges.

If the neural network has N neurons, the ith neuron determines its output value as follows:

1. Compute a weighted sum of the output values of its neighbors. This is given by $\sum_{j=1}^{N} T_{ij} V_j$.

2. Compare the weighted sum with its threshold I_i. The output value

V_i of the neuron is given by the following *update rule*:

$$V_i = \begin{cases} 1 & \text{if } \sum_{j=1}^{N} T_{ij}V_j + I_i > 0 \\ 0 & \text{if } \sum_{j=1}^{N} T_{ij}V_j + I_i < 0 \\ V_i & \text{otherwise.} \end{cases} \quad (4.1)$$

Hopfield [1] showed that the computation performed by a network of such elements is equivalent to finding a minimum of the following energy function:

$$E = -\frac{1}{2}\sum_{i=1}^{N}\sum_{j=1}^{N} T_{ij}V_iV_j - \sum_{i=1}^{N} I_iV_i \quad (4.2)$$

To understand this equivalence, consider the change E in E due to an incremental change V_i in the state of neuron i. The incremental change in energy, as derived from Equation 4.2, is given by:

$$E = -[\sum_{j=1}^{N} T_{ij}V_j + I_i]V_i \quad (4.3)$$

Since the update rule (Equation 4.1) guarantees that V_i is positive (negative) only when the quantity in the square brackets is positive (negative), any change in E due to a neuron state-change will always be negative and the computation performed by the neural network can be interpreted as minimization of E.

To solve combinatorial optimization problems with neural networks, one constructs a suitable energy function with minima that can be interpreted as solutions to the given problem. From the energy function, we determine the interconnections, respective link-weights, and thresholds of all neurons. Starting in any state, if we allow the neurons to change states, the network will finally settle at a minimum of the energy function.

4.2 Electrical Neural Networks

Figure 4.3 shows a possible implementation of a neuron using analog circuit components. It consists of two high-gain amplifiers with gain transfer functions g_i and $-g_i$. The input voltage to the neuron is denoted by u_i, and V_i is the magnitude of the output voltage. Consider a fully-connected network of N neurons as shown in Figure 4.4. Resistors (indicated by solid dots in Figure 4.4) connect the output of an amplifier to the inputs of other amplifiers. The time evolution (behavior) of such a network is completely

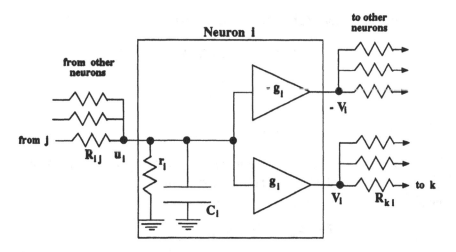

FIGURE 4.3: An analog computing element.

described by a set of N first-order differential equations, each expressing the rate of change of the input voltage of an amplifier in terms of the output voltages of all other amplifiers. From elementary circuit theory, if r_i and C_i are the input resistance and input capacitance of the two amplifiers of neuron i (see Figure 4.3), respectively, and R_{ij} is the resistor connecting the output of neuron j to the input line of neuron i, then the input voltage u_i and the output voltage V_i are related as follows:

$$C_i \frac{du_i}{dt} = \sum_{j=1}^{N} \frac{(V_j - u_i)}{R_{ij}} - \frac{u_i}{r_i} + I_i \qquad (4.4)$$

where I_i is a fixed input current flowing into the amplifiers of the ith neuron (I_i is also referred to as the threshold of neuron i). We assume that the output impedance of all amplifiers is negligible and substitute

$$\frac{1}{R_i} = \frac{1}{r_i} + \sum_{j=1}^{N} \frac{1}{R_{ij}} \; , \quad \tau_i = R_i C_i \qquad (4.5)$$

Also, we adjust the r_i's so that all amplifiers have the same time constant τ. We can then replace $\frac{1}{R_{ij}C_i}$ by T_{ij}, redefine $\frac{I_i}{C_i}$ as I_i, and rearrange Equation 4.4 to obtain:

$$\frac{du_i}{dt} = \sum_{j=1}^{N} T_{ij}V_j - \frac{u_i}{\tau} + I_i \qquad (4.6)$$

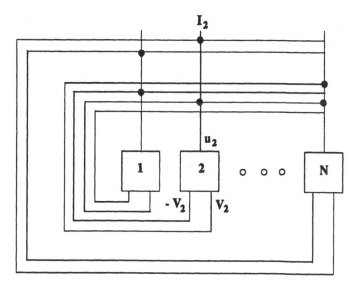

FIGURE 4.4: An analog neural network.

Hopfield [1] has shown that the time evolution of such a network is identical to the minimization of the following function:

$$E = -\frac{1}{2}\sum_{i=1}^{N}\sum_{j=1}^{N} T_{ij}V_iV_j + \frac{1}{\lambda\tau}\sum_{i=1}^{N}[\int_0^{V_i} g_i^{-1}(V)\,dV] - \sum_{i=1}^{N} I_iV_i \quad (4.7)$$

where the input-output relationship of the ith amplifier is given by $V = g_i(\lambda u)$. Here, u and V are the input and output voltages, respectively, of the amplifier and $\lambda \geq 0$ is a constant that can be used to change the gain of the amplifier. Let us take the derivative of E with respect to time t. Thus,

$$\frac{dE}{dt} = \sum_{i=1}^{N} \frac{dE}{dV_i}\frac{dV_i}{dt}$$

$$= -\sum_{i=1}^{N}(\sum_{j=1}^{N} T_{ij}V_j - \frac{u_i}{\tau} + I_i)\frac{dV_i}{dt}$$

Using Equation 4.6,

$$\frac{dE}{dt} = -\sum_{i=1}^{N} \frac{du_i}{dt}\frac{dV_i}{dt}$$

$$= -\frac{1}{\lambda}\sum_{i=1}^{N} \frac{d(g_i^{-1}(V_i))}{dV_i}(\frac{dV_i}{dt})^2$$

If we restrict $g_i^{-1}(V_i)$ to be a monotone increasing function (which usually is the case with the transfer function of the amplifier), each term in the sum is non-negative. Therefore, $\frac{dE}{dt} \leq 0$, and $\frac{dE}{dt} = 0$ implies $\frac{dV_i}{dt} = 0$. When the energy function cannot be decreased any further, the neuron outputs cease to change and the neural network reaches an equilibrium state. Starting in any initial state of the neurons, this network will go through a succession of state changes to eventually settle in some equilibrium state. This is known as the *time evolution* of the network state.

Notice that in the high-gain limit, $\lambda \rightarrow \infty$, the integral term in Equation 4.7 becomes negligible and the energy function can be approximated as Equation 4.2. Basically, amplifiers with very high gain behave as two-state devices. Therefore, discrete combinatorial optimization problems formulated as the minimization of Equation 4.2 can be solved using a network of high-gain amplifiers.

Neural networks are typically used for two types of applications [4]. In the first application, the network performs constraint satisfaction tasks involving multiple constraints (combinatorial optimization) to find a consistent set of states. In the second application, the neural network learns from examples. The essential difference between the two applications is as follows: in the first application the network adjusts the states of the neurons to a given fixed set of connection strengths, whereas in the second application the network adjusts the values of the connection strengths to given example sets of neuron states. In the second application, methods such as the *backpropagation* technique are used to learn from examples.

Our present application belongs to the first category. We formulate the testing problem as a neural network optimization problem (Chapter 6).

References

[1] J. J. Hopfield. Neurons with Graded Response Have Collective Computational Properties Like Those of Two State Neurons. *Proceedings of the National Academy of Sciences*, 81(10):3088–3092, May 1984.

[2] J. J. Hopfield. Artificial Neural Networks. *IEEE Circuits and Devices Magazine*, 4(5):3–10, September 1988.

[3] R. L. Lippmann. An Introduction to Computing with Neural Nets. *IEEE Acoustics, Speech and Signal Processing Magazine*, 4:4–22, 1987.

[4] D. E. Rumelhart and J. L. McClelland. *Parallel Distributed Processing: Explorations in the Microstructure of Cognition*. MIT Press, Cambridge, Massachusetts, 1987.

Chapter 5

NEURAL MODELING FOR DIGITAL CIRCUITS

> *"Science has little use for models that slavishly obey all our wishes. We want models that talk back to us, models that have a mind of their own. We want to get out of our models more than we have put in. A reasonable way to start is to put in as little as possible."*
>
> – T. Toffoli and N. Margolus in *Cellular Automata Machines*, MIT Press (1987)

We can relate the input and output signal states of a logic gate through an *energy function*, defined over a network of neurons, such that the minimum-energy states correspond to the gate's function. Similarly, the function of an entire circuit, consisting of any arbitrary interconnections of logic gates, can be expressed by a single energy function. Several reasons motivate this new approach. First, since the function of the circuit is captured in the energy expression, mathematical techniques such as non-linear programming can be used to solve test generation and other design problems. Second, graph-theoretic techniques can be applied to the neural network graph to accelerate the minimization of the energy function. Third, the non-causal form of this model allows the use of parallel processing for compute-intensive design automation tasks.

33

5.1   Logic Circuit Model

Every net (signal) in the circuit is represented by a neuron and the value on the net is the activation value (0 or 1) of the neuron. The neurons corresponding to the primary inputs (outputs) of the circuit are called *primary input (output) neurons*. Neurons corresponding to the input (output) nets of a gate are called *input (output) neurons*. Each gate is independently mapped onto a neural network and the interconnections between the gates are used to combine the individual gate neural networks into a neural network representing the circuit. Neural networks for 2-input AND, OR, NAND, NOR, XOR and XNOR gates and the single-input inverter (NOT gate) constitute the *basis set*. Any gates with more than two inputs are constructed from this basis set [1, 2].

The neural network for a digital circuit is characterized by an energy function E that has global minima only at the neuron states consistent with the functions of all gates in the circuit. All other neuron states have higher energy. The energy function E assumes a value Z for all *consistent states* and $E > Z$ for all inconsistent states. Also, the value Z is known *a priori* so that we can easily recognize a global minimum. A consistent labeling of signals in the circuit can be obtained by finding a *minimizing point* of the energy function, that is, a set of values of variables V_i's for which $E = Z$.

The energy function E is uniquely specified by the weights on the links connecting neurons and the thresholds of the neurons. The weights are expressed as a matrix T and the thresholds, as a vector I. The energy function, as given by Hopfield [4], is

$$E = -\frac{1}{2}\sum_{i=1}^{N}\sum_{j=1}^{N}T_{ij}V_iV_j - \sum_{i=1}^{N}I_iV_i + K \qquad (5.1)$$

where N is the number of neurons in the neural network, $T_{ij} = T_{ji}$ is the weight of the link between neurons i and j, V_i is the activation value of neuron i, I_i is the threshold of neuron i and K is a constant. Also, $T_{ii} = 0$.

The energy function E in Equation 5.1 is obtained by appropriately summing the energy functions for all primitive gates of the circuit. The energy function E_G associated with the neural network for a primitive gate G is of the same form as E such that $E_G = Z_G$ for all neuron states consistent with the function of the gate and $E_G > Z_G$ for all other neuron states (Z_G is some constant). A formal derivation of the parameters of the energy functions for the basis set of logic gates is given in Section 5.2.2. Table 5.1 shows T and I for the neural networks in the basis set in terms

TABLE 5.1: Neural network parameters for logic gates.

Gate	T and I
AND	$T_{ij} = (1 - \delta(i,j)) \times ((A + B) \times connected(i,j)$ $\qquad\qquad - B \times inputs(i,j))$ $I_i = -(2A + B) \times output(i)$ $K = 0$
OR	$T_{ij} = (1 - \delta(i,j)) \times ((A + B) \times connected(i,j)$ $\qquad\qquad - B \times inputs(i,j))$ $I_i = -B \times output(i) - A \times input(i)$ $K = 0$
NAND	$T_{ij} = (1 - \delta(i,j)) \times ((-A - B) \times connected(i,j)$ $\qquad\qquad - B \times inputs(i,j))$ $I_i = (2A + B) \times output(i) + (A + B) \times input(i)$ $K = 2A + B$
NOR	$T_{ij} = (1 - \delta(i,j)) \times ((-A - B) \times connected(i,j)$ $\qquad\qquad - B \times inputs(i,j))$ $I_i = B$ $K = B$
NOT	$T_{ij} = -2J \times (1 - \delta(i,j))$ $I_i = J$ $K = J$

of the following functions:

$$inputs(i,j) = \begin{cases} 1 & \text{if } i \text{ and } j \text{ are} \\ & \text{input neurons for the same gate.} \\ 0 & \text{otherwise.} \end{cases}$$

$$connected(i,j) = 1 - inputs(i,j)$$

$$\delta(i,j) = \begin{cases} 1 & \text{if } i = j. \\ 0 & \text{otherwise.} \end{cases}$$

$$output(i) = \begin{cases} 1 & \text{if } i \text{ is an output neuron for the gate.} \\ 0 & \text{otherwise.} \end{cases}$$

$$input(i) = \begin{cases} 1 & \text{if } i \text{ is an input neuron for the gate.} \\ 0 & \text{otherwise.} \end{cases}$$

In Table 5.1, A, B, C and J are constants such that $A > 0$, $B > 0$, $C > 0$, $J > 0$ and $B > C$. Note that the energy function E_G for a network in the basis set assumes a global minimum value 0 at all consistent network states and E_G is greater than this minimum value for all other network states. It is not essential for E_G to have a minimum value of 0 for any specific network G in the basis set. In fact, by adding the constant term K, the function E_G can be made to have any arbitrary value at its global minimum. Without loss of generality, in the sequel we assume that the energy function E_G for any gate in the basis set has a minimum value of 0.

Neural networks for two-input AND, OR, NAND and NOR gates consist of three neurons that correspond to the input and output nets of the gate. As an illustration, Figure 5.1a shows the neural network for an AND

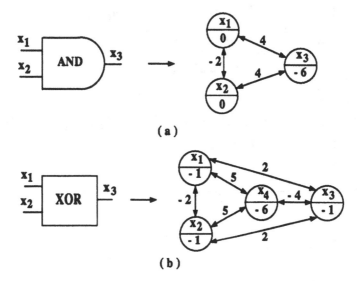

(a)

(b)

FIGURE 5.1: Networks for AND (a) and XOR (b) gates with $A = B = 2$, $C = 1$.

gate. The three neurons are labeled x_1, x_2 and x_3. Each circle corresponds to a neuron. The name of the neuron is written in the upper half and its threshold is indicated in the lower half. Let V_i and I_i denote the activation value and the threshold of neuron x_i, and T_{ij} denote the link weight between neurons x_i and x_j. The energy function for the AND gate, from Equation 5.1, is

$$E_{AND} = -\frac{1}{2}\sum_{i=1}^{3}\sum_{j=1}^{3}T_{ij}V_iV_j - \sum_{i=1}^{3}I_i V_i + K$$

where T_{ij}, I_i and K are given in Table 5.1. E_{AND} is 0 at all four consistent states $(V_1 = V_2 = V_3 = 0)$, $(V_1 = V_2 = V_3 = 1)$, $(V_1 = 1, V_2 = V_3 = 0)$ and $(V_1 = V_3 = 0, V_2 = 1)$ corresponding to the truth table of the AND gate. $E_{AND} > 0$ for the all other states that are inconsistent with the

TABLE 5.2: Energy surface for E_{AND}.

AND			
V_1	V_2	V_3	E_{AND}
0	0	0	0
0	0	1	$2A + B$
0	1	0	0
0	1	1	A
1	0	0	0
1	0	1	A
1	1	0	B
1	1	1	0

function of the AND gate. Table 5.2 shows the value of the energy function for all possible states of the network.

The XOR (Figure 5.1b) and XNOR gate neural networks have an additional neuron that does not correspond to any external net associated with the gate (see Table 5.3). This neuron has an activation value 1 when both inputs to the gate are 1. Section 5.2 gives a mathematical basis for these transformations.

5.2 Existence of Neural Models

We show that an arbitrary logic circuit can be represented by a neural network. The summation of energy functions for the individual gates yields the energy function for the logic circuit. Since the individual gate energy functions only assume non-negative values, the energy function for the circuit is also non-negative and it is 0 when each gate energy function separately becomes 0. As an illustration, consider the logic circuit shown in Figure 5.2a. The neural networks for the two AND gates are shown in Figure 5.2b. Note that there are duplicate neurons with labels x_2 and x_3 since the nets x_2 and x_3 are connected to both AND gates. Similarly, the link between x_2 and x_3 appears twice with two different values of T_{23}. The individual neural networks for the two AND gates are merged to yield the neural network for the logic circuit (see Figure 5.2c). We collapse neurons with identical labels by a single neuron having a threshold that is the sum of the individual thresholds. Similarly, we merge the links between the collapsed neurons into a single link with link-weight equal to the sum of the individual link-weights. Thus, we replace the two neurons labeled x_2 in Figure 5.2b by a single neuron x_2 with a threshold of 0, the two neurons la-

TABLE 5.3: Neural network parameters for XOR and XNOR gates.

XOR	$T = \begin{pmatrix} 0 & -B & 2C & (A+B+C) \\ -B & 0 & 2C & (A+B+C) \\ 2C & 2C & 0 & -4C \\ (A+B+C) & (A+B+C) & -4C & 0 \end{pmatrix}$ $I = \begin{pmatrix} -C & -C & -C & -(2A+B) \end{pmatrix}$ $K = 0$
XNOR	$T = \begin{pmatrix} 0 & -B & -2C & (A+B+C) \\ -B & 0 & -2C & (A+B+C) \\ -2C & -2C & 0 & 4C \\ (A+B+C) & (A+B+C) & 4C & 0 \end{pmatrix}$ $I = \begin{pmatrix} C & C & C & -(2A+B+4C) \end{pmatrix}$ $K = C$

beled x_3 by a single neuron x_3 with threshold -6 and the two links labeled 4 and -2 by a single link T_{23} of weight 2. This completes the construction of the neural network for the example circuit in Figure 5.2a.

5.2.1 Neural Networks In Basis Set are Optimal

An N-terminal device has at least N neurons in its neural network model. A neural network for a gate is called *optimal* if no other network representing the gate exists with fewer neurons. Three-neuron symmetrically-connected networks exist for 2-input AND, OR, NAND and NOR gates and are optimal since the gates have three terminals. However, not all 2-input single output Boolean functions can be represented by 3-neuron networks. For example, 2-input XOR or XNOR gates can only be represented by 4-neuron networks (Figure 5.1b) which are optimal. In the following, we give a *necessary* condition for the existence of an N-neuron network for an arbitrary N-terminal device. The 3-terminal XOR and XNOR gates do not satisfy the necessary condition and hence require a 4-neuron network.

Definition: Associated with each neuron i is a hyperplane $I_i + \sum_{j=1}^{N} T_{ji} V_j = 0$ in an $N - 1$ dimensional space and a point p in this space is a set of activation values of all neurons except i. Also, associated with neuron

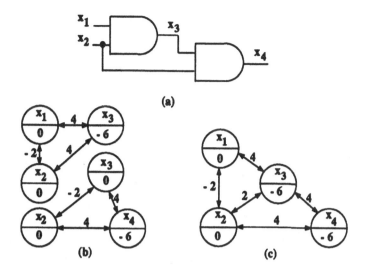

FIGURE 5.2: A circuit and the corresponding neural network ($A = B = 2$).

i there are three sets, P_{i_on}, P_{i_off} and P_{i_other}, whose elements are the points corresponding to consistent states of the network. A point belongs to P_{i_on} (P_{i_off}) if it corresponds to only one consistent state and the neuron i has an activation value 1 (0). P_{i_other} consists of all other points that correspond to consistent states but are not in the sets P_{i_on} and P_{i_off}.

Example 5.1: The sets associated with neuron 3 in the AND gate in Figure 5.1a are $P_{3\_on} = \{(V_1 = 1, V_2 = 1)\}$, $P_{3\_off} = \{(V_1 = 0, V_2 = 0), (V_1 = 1, V_2 = 0), (V_1 = 0, V_2 = 1)\}$, and $P_{3\_other} = \{\}$. The sets associated with neuron 2 are $P_{2\_on} = \{(V_1 = 1, V_3 = 1)\}$, $P_{2\_off} = \{(V_1 = 1, V_3 = 0)\}$ and $P_{2\_other} = \{(V_1 = 0, V_3 = 0)\}$. The point $(V_1 = 0, V_3 = 1)$ corresponds to inconsistent states of the network and, therefore, is not included in P_{2_on}, P_{2_off}, or P_{2_other}.

Definition: A hyperplane $I_i + \sum_{j=1}^{N} T_{ji}V_j = 0$ associated with neuron i is a *decision hyperplane* if the points in P_{i_on} and P_{i_off} fall on opposite sides of the hyperplane and all points in P_{i_other} lie on the hyperplane.

Example 5.2: The hyperplane associated with neuron 2 in Figure 5.1a is $4V_3 - 2V_1 = 0$ since $I_2 = 0$, $T_{23} = 4$ and $T_{12} = -2$. Figure 5.3 shows the hyperplane. Sets P_{2_on}, P_{2_off} and P_{2_other} are as computed in Example 5.1. Clearly, points in P_{2_on} and P_{2_off} fall on opposite sides of the hyperplane and the point in P_{2_other} lies on the hyperplane.

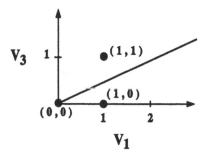

FIGURE 5.3: Hyperplane for neuron 2 in Figure 5.1a.

Theorem 5.1: *A necessary condition for a network of N neurons and an energy function E defined in Equation 5.1 to model a device with N terminals is the existence of a decision hyperplane for each of the N neurons.*

Proof: The difference between the global energy of the network when neuron i is off and when it is on, given the current states of all other neurons, is derived from Equation 5.1 as follows:

$$E(V_i = 0) - E(V_i = 1) \quad = \quad E_i = I_i + \sum_{j=1}^{N} T_{ji} V_j$$

For an arbitrary point $p \in P_{i\_on}$ (P_{i_off}) neuron i has an activation value 1 (0) in the consistent state S_1 and 0 (1) in the inconsistent state S_2. The energy function E should have a lower value of energy for the consistent state S_1 as compared to the inconsistent state S_2. Therefore, E_i should necessarily be positive (negative) and the decision hyperplane $E_i = 0$ divides the $N - 1$ dimensional space into two regions R1 $(E_i > 0)$ and R2 $(E_i < 0)$. For E to be zero only at consistent states, all points in P_{i_on} (P_{i_off}) must lie in region R1 (R2).

For any arbitrary point $p \in P_{i\_other}$, there are two consistent states S_1 and S_2 with neuron i having activation values 0 and 1, respectively. Since E should attain its minimum value 0 in both of these states, it is mandatory that E_i be zero. Hence, p must lie on the decision hyperplane. Thus, for any given neuron i, the existence of a decision hyperplane is a necessary condition for the existence of the neural model.   ∎

5.2.2 Parameters of Energy Function

We present a formal derivation of the energy function for the 3-neuron neural network model of the AND gate in Figure 5.1a. Without loss of generality, let the energy function assume the value 0 at all consistent states of the AND gate and a value greater than 0 at all other states. It can be graphically verified that decision hyperplanes exist for all three neurons. The energy function

$$E_{AND} = -\frac{1}{2}\sum_{i=1}^{3}\sum_{j=1}^{3}T_{ij}V_iV_j - \sum_{i=1}^{3}I_i V_i + K \qquad (5.2)$$

should be 0 at all four consistent states, $(V_1 = V_2 = V_3 = 0)$, $(V_1 = V_2 = V_3 = 1)$, $(V_1 = 1, V_2 = V_3 = 0)$ and $(V_1 = V_3 = 0, V_2 = 1)$, corresponding to the truth table of the AND gate. Substituting $(V_1 = V_2 = V_3 = 0)$ in Equation 5.2, we get $K = 0$. Similarly, substituting $(V_1 = V_3 = 0, V_2 = 1)$ and $(V_1 = 1, V_2 = V_3 = 0)$ in Equation 5.2, we get: $I_2 = 0$ and $I_1 = 0$, respectively. Substituting $(V_1 = V_2 = V_3 = 1)$ in Equation 5.2, we get

$$\frac{1}{2}\sum_{i=1}^{3}\sum_{j=1}^{3}T_{ij} + I_3 = 0$$

The existence of a decision hyperplane for neuron 1 requires that $T_{23} + T_{13} + I_3 > 0$ and $I_3 < 0$. Similarly, a decision hyperplane for neuron 2 requires that $T_{12} < 0$. The decision hyperplane for neuron 1 does not have any additional constraint. Had there been inconsistent states at a Hamming distance > 1 from any consistent state, the condition $E > 0$ would require additional inequalities. A set of values satisfying the above equalities and inequalities are $T_{23} = T_{13} = A + B$, $T_{12} = -B$, $I_3 = -(2A + B)$ and $I_1 = I_2 = K = 0$, where A and B are any positive constants. This technique also provides the energy functions for the neural network models of other Boolean gates in the basis set. An alternative derivation, however, is presented in Section 5.3.

5.3 Properties of Neural Models

We discuss some general properties of the neural network models. These properties are used in the *neural pruning* stage of the ATG system (Section 7.2) to minimize the number of neurons needed to represent a logic circuit. The same properties can also be used to derive the energy functions for all networks in the basis set, given the energy function of any one gate in the basis set.

We consider a class of $(N - 1)$-input single-output switching functions

$$V_N = f(V_1, \ldots, V_{N-1}) \qquad (5.3)$$

that can be represented by a N-neuron network $H(V_1, \ldots, V_{N-1}, V_N)$, where V_i is the activation value of the ith neuron. The results of this section are easily extended to an $(N-1)$-input switching function represented by more than N neurons (e.g., XOR gates). The structure of H is given by $[T; I; K]$ where T is the $N \times N$ weight matrix, I is the vector of thresholds and K is a constant associated with E_H, the energy function of H. $T_{ij} = T_{ji}$ is the weight on the link connecting neurons i and j, $T_{ii} = 0$ and I_j is the threshold of neuron j. Without loss of generality, we assume that $E_H = 0$ at all consistent states of the network.

Given the switching function of Equation 5.3, a *single-variable inversion* (SVI) function f^i is defined as follows:

$$f^i(V_1, \ldots, V_i, \ldots, V_{N-1}) = \begin{cases} f(V_1, \ldots, \overline{V}_i, \ldots, V_{N-1}) & : \quad i \neq N \\ \overline{f}(V_1, \ldots, V_i, \ldots, V_{N-1}) & : \quad i = N \end{cases}$$

$$(5.4)$$

For example, Figure 5.4a shows the function f, Figure 5.4b shows the SVI

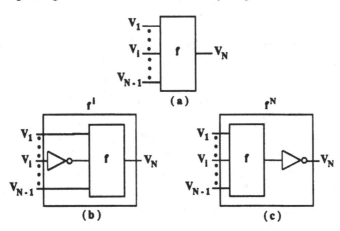

FIGURE 5.4: Functions f (a), f^i (b) and f^N (c).

function f^i ($i \neq N$) and Figure 5.4c shows the SVI function f^N. Notice, f^N is the negation of function f. Let $H'(V_1, \ldots, V_{N-1}, V_N)$ be the neural network corresponding to a SVI function and let $[T'; I'; K']$ be the structure of H'.

Theorem 5.2: *If $H = [T; I; K]$ is a neural network for function f, then the SVI function f^p ($1 \leq p \leq N$) is realized by the network $H' = [T'; I'; K']$, where*

1. $I'_p = -I_p$. For $j \neq p$, $I'_j = I_j + T_{jp}$, i.e., thresholds of all neurons $j \neq p$ in H are increased by T_{jp}.

2. $T'_{jp} = T'_{pj} = -T_{pj}$ for $1 \leq j \leq N$, i.e., weights on all links connected to neuron p are multiplied by -1. For $i \neq p$ and $j \neq p$, $T'_{ij} = T_{ij}$.

3. $K' = K - I_p$.

Proof: Unless stated otherwise, $1 \leq j \leq N$ and $1 \leq i \leq N$. Let S and S' be the set of consistent states of networks H and H', respectively. Let P_{i_on}, P_{i_off} and P_{i_other} be the sets associated with neuron i in network H and let P'_{i_on}, P'_{i_off} and P'_{i_other} be the corresponding sets in network H'.

The proof is based on the following observation (see Figure 5.4): If a binary vector[1] $v_1, \ldots, v_p, \ldots, v_N$ satisfies

$$v_N = f(v_1, \ldots, v_p, \ldots, v_{N-1})$$

then the vector $v_1, \ldots, 1 - v_p, \ldots, v_N$ satisfies

$$v_N = f^p(v_1, \ldots, 1 - v_p, \ldots, v_{N-1}).$$

Therefore, if $v_1, \ldots, v_p, \ldots, v_N$ is a consistent state of the network H then the vector $v_1, \ldots, 1 - v_p, \ldots, v_N$ must be a consistent state of the network H'. Hence, set S' can easily be derived from S and the consistent states of networks H and H' are related as follows:

1. $P'_{p\_on} = P_{p\_off}$, $P'_{p\_off} = P_{p\_on}$ and $P'_{p\_other} = P_{p\_other}$.

2. For a neuron $j \neq p$, P'_{j_on}, P'_{j_off} and P'_{j_other} are the sets obtained by substituting $1 - v_p$ for v_p in every point in P_{j_on}, P_{j_off} and P_{j_other}, respectively. For example, if point

$$(v_1, \ldots, v_{j-1}, v_{j+1}, \ldots, v_p, \ldots, v_N) \in P_{j\_on}$$

then

$$(v_1, \ldots, v_{j-1}, v_{j+1}, \ldots, 1 - v_p, \ldots, v_N) \in P'_{j\_on}.$$

3. If $E_H(v_1, \ldots, v_p, \ldots, v_N) = 0$ then $E_{H'}(v_1, \ldots, 1 - v_p, \ldots, v_N) = 0$.

In order to find the values of I'_j and T'_{ij} so that neurons in network H' satisfy the above conditions, we consider the following cases.

---

[1] In this discussion, v_i is used to denote a specific value of the variable V_i.

CASE 1:

Consider neuron p. From Section 5.2.1, points in P_{p_on}, P_{p_off} and P_{p_other} satisfy inequalities (5.5) and (5.6) and Equation 5.7, respectively, as follows:

$$T_{p1}v_1 + T_{p2}v_2 + \cdots + T_{pN}v_N + I_p \; > \; 0 \tag{5.5}$$
$$T_{p1}v_1 + T_{p2}v_2 + \cdots + T_{pN}v_N + I_p \; < \; 0 \tag{5.6}$$
$$T_{p1}v_1 + T_{p2}v_2 + \cdots + T_{pN}v_N + I_p \; = \; 0 \tag{5.7}$$

If we substitute

$$I'_p \; = \; -I_p$$
$$T'_{pj} \; = \; -T_{pj} : 1 \leq j \leq N$$

we get

$$T'_{p1}v_1 + T'_{p2}v_2 + \cdots + T'_{pN}v_N + I'_p \; < \; 0 \tag{5.8}$$
$$T'_{p1}v_1 + T'_{p2}v_2 + \cdots + T'_{pN}v_N + I'_p \; > \; 0 \tag{5.9}$$
$$T'_{p1}v_1 + T'_{p2}v_2 + \cdots + T'_{pN}v_N + I'_p \; = \; 0 \tag{5.10}$$

Inequality (5.8) is only satisfied by points in P'_{p_off}, inequality (5.9) is only satisfied by points in P'_{p_on} and Equation 5.10 is only satisfied by points in P'_{p_other}. Thus, for the above choice of T'_{pj} and I'_p, $P_{p\_on} = P'_{p\_off}$, $P_{p\_off} = P'_{p\_on}$ and $P_{p\_other} = P'_{p\_other}$.

CASE 2:

Consider a neuron $j \neq p$. Points in P_{j_on}, P_{j_off} and P_{j_other} satisfy inequalities (5.11) and (5.12), and Equation 5.13, respectively, as given below:

$$T_{j1}v_1 + T_{j2}v_2 + \cdots + T_{jp}v_p + \cdots + T_{jN}v_N + I_j \; > \; 0 \tag{5.11}$$
$$T_{j1}v_1 + T_{j2}v_2 + \cdots + T_{jp}v_p + \cdots + T_{jN}v_N + I_j \; < \; 0 \tag{5.12}$$
$$T_{j1}v_1 + T_{j2}v_2 + \cdots + T_{jp}v_p + \cdots + T_{jN}v_N + I_j \; = \; 0 \tag{5.13}$$

By substituting as follows,

$$T'_{jp} \; = \; -T_{jp}$$
$$T'_{ji} \; = \; T'_{ij} = T_{ji} : 1 \leq i \leq N, \, i \neq p$$
$$I'_j \; = \; I_j + T_{jp}$$

we get

$$T'_{j1}v_1 + T'_{j2}v_2 + \cdots + T'_{jp}(1 - v_p) + \cdots + T'_{jN}v_N + I'_j \; > \; 0 \tag{5.14}$$

$$T'_{j1}v_1 + T'_{j2}v_2 + \cdots + T'_{jp}(1 - v_p) + \cdots + T'_{jN}v_N + I'_j \; < \; 0 \tag{5.15}$$

$$T'_{j1}v_1 + T'_{j2}v_2 + \cdots + T'_{jp}(1 - v_p) + \cdots + T'_{jN}v_N + I'_j \; = \; 0 \tag{5.16}$$

Inequality (5.14) is only satisfied by points in P'_{j_on}, inequality (5.15) is only satisfied by points in P'_{j_off} and Equation 5.16 is only satisfied by points in P'_{j_other}. Clearly, points satisfying the first (second) inequality differ from the points in P_{j_on} (P_{j_off}) only in the value of neuron p. Similarly, points satisfying Equation 5.16 differ from the points in P_{j_other} only in the value of neuron p.

CASE 3:

Consider a consistent state $v_1, \ldots, v_p, \ldots, v_N$ of neural network H. From Equation 5.1,

$$E_H = -\frac{1}{2}\sum_{i=1}^{N}\sum_{j=1}^{N}T_{ij}v_iv_j - \sum_{i=1}^{N}I_iv_i + K = 0 \tag{5.17}$$

Also, from Equation 5.1,

$$
\begin{aligned}
E_{H'} &= -\frac{1}{2}\sum_{i=1}^{N}\sum_{j=1}^{N}T'_{ij}v_iv_j - \sum_{i=1}^{N}I'_iv_i + K' \\
&= -\frac{1}{2}[\sum_{\substack{i=1 \\ i\neq p}}^{N}\sum_{\substack{j=1 \\ j\neq p}}^{N}T'_{ij}v_iv_j + 2\sum_{j=1}^{N}T'_{jp}v_jv_p] - [\sum_{\substack{i=1 \\ i\neq p}}^{N}I'_iv_i + I'_pv_p] + K'
\end{aligned}
$$

Substituting the values of T'_{ij} and I'_j as determined in CASE 1 and CASE 2, we get:

$$
\begin{aligned}
E_{H'} &= -\frac{1}{2}[\sum_{\substack{i=1 \\ i\neq p}}^{N}\sum_{\substack{j=1 \\ j\neq p}}^{N}T_{ij}v_iv_j - 2\sum_{j=1}^{N}T_{jp}v_jv_p] \\
&\quad - [\sum_{\substack{i=1 \\ i\neq p}}^{N}(I_i + T_{ip})v_i - I_pv_p] + K'
\end{aligned}
$$

For $v_1, \ldots, 1 - v_p, \ldots, v_N$ to be a consistent state of the network H', we must have $E_{H'}(v_1, \ldots, 1 - v_p, \ldots, v_N) = 0$. Therefore,

$$
\begin{aligned}
E_{H'} &= -\frac{1}{2}[\sum_{\substack{i=1 \\ i \neq p}}^{N} \sum_{\substack{j=1 \\ j \neq p}}^{N} T_{ij} v_i v_j - 2 \sum_{j=1}^{N} T_{jp} v_j + 2 \sum_{j=1}^{N} T_{jp} v_j v_p] \\
&\quad - [\sum_{\substack{i=1 \\ i \neq p}}^{N} (I_i + T_{ip}) v_i - I_p(1 - v_p)] + K' \\
&= -\frac{1}{2} \sum_{i=1}^{N} \sum_{j=1}^{N} T_{ij} v_i v_j - \sum_{i=1}^{N} I_i v_i + I_p + K' \\
&= -K + I_p + K' \\
&= 0
\end{aligned}
$$

Hence, $K' = K - I_p$ and $H' = [T'; I'; K']$. Therefore, the neural network for the SVI function f^p can be obtained from the network for f as claimed.

∎

Example 5.3: An AND gate realizes the switching function

$$V_3 = f(V_1, V_2) = V_1 \wedge V_2$$

and $f(\overline{V}_1, \overline{V}_2) = \overline{V}_1 \wedge \overline{V}_2$ is realized by a NOR gate. The neural network for an AND gate can be transformed into a neural network for the NOR gate, by applying Theorem 5.2 to each of the inputs of the AND gate. From Table 5.1, setting $A = B = 2$, the AND gate neural network $H = [T; I; K]$ is

$$
T = \begin{pmatrix} 0 & -2 & 4 \\ -2 & 0 & 4 \\ 4 & 4 & 0 \end{pmatrix} \qquad I = \begin{pmatrix} 0 \\ 0 \\ -6 \end{pmatrix} \qquad K = 0
$$

Applying Theorem 5.2 for $V_p = V_2$, we get $T'_{23} = T'_{32} = -T_{23} = -4$, $T'_{12} = T'_{21} = -T_{12} = 2$, $I'_3 = I_3 + T_{23} = -2$, $I'_2 = -I_2 = 0$ and $I'_1 = I_1 + T_{12} = -2$. Thus, $f(V_1, \overline{V}_2)$ is realized by a neural network $H' = [T'; I'; K']$ given by

$$
T' = \begin{pmatrix} 0 & 2 & 4 \\ 2 & 0 & -4 \\ 4 & -4 & 0 \end{pmatrix} \qquad I' = \begin{pmatrix} -2 \\ 0 \\ -2 \end{pmatrix} \qquad K' = 0
$$

Now applying Theorem 5.2 to H' for $V_p = V_3$, the neural network $H'' = [T''; I''; K'']$ for $f(\overline{V}_1, \overline{V}_2)$ is given by

$$T'' = \begin{pmatrix} 0 & -2 & -4 \\ -2 & 0 & -4 \\ -4 & -4 & 0 \end{pmatrix} \qquad I'' = \begin{pmatrix} 2 \\ 2 \\ 2 \end{pmatrix} \qquad K'' = 2$$

From Table 5.1, it can easily be verified that T'', I'' and K'' are the weight matrix, threshold vector and constant associated with a NOR gate with V_1 and V_2 as inputs and V_3 as the output, when $A = B = 2$.

Corollary 5.1: *If the neural network for $V_N = f(V_1, V_2, \ldots, V_{N-1})$ is given by $H = [T; I; K]$, its inversion $\overline{V}_N = \overline{f}(V_1, V_2, \ldots, V_{N-1})$ is then realized by $H' = [T'; I'; K']$, where*

1. $I'_N = -I_N$. *For* $j \neq N$, $I'_j = I_j + T_{jN}$.

2. $T'_{jN} = T'_{Nj} = -T_{jN}$. *For* $i \neq N, j \neq N, T'_{ij} = T'_{ji} = T_{ij}$.

3. $K' = K - I_N$.

The dual of function f is the function $\overline{f}(\overline{V}_1, \ldots, \overline{V}_{N-1})$. Therefore, starting from f, we can construct a sequence of functions $\mathcal{F}_i$, $1 \leq i \leq N$, such that $\mathcal{F}_1$ is the SVI function f^1 and $\mathcal{F}_i$, $i \neq 1$, is the SVI function $\mathcal{F}^i_{i-1}$. Clearly, $\mathcal{F}_N$ is the dual of f.

Example 5.4: If $V_3 = f(V_1, V_2)$, then $\mathcal{F}_1 = f(\overline{V}_1, V_2)$. $\mathcal{F}_2$ is the SVI function $\mathcal{F}^2_1$ which is $f(\overline{V}_1, \overline{V}_2)$. Similarly, $\mathcal{F}_3 = \overline{f}(\overline{V}_1, \overline{V}_2)$ which is the dual of f.

Theorem 5.3: *If a neural network for $V_N = f(V_1, \ldots, V_{N-1})$ has a structure $H = [T; I; K]$, the dual of f is realized by a network $H' = [T'; I'; K']$, where*

$$I'_i = -I_i - \sum_{j=1}^{N} T_{ij} \qquad (5.18)$$

$$T'_{ij} = T_{ij} \qquad (5.19)$$

$$K' = K - \sum_{i=1}^{N} I_i - \frac{1}{2}\sum_{i=1}^{N}\sum_{j=1}^{N} T_{ij} \qquad (5.20)$$

Proof: We repeatedly use Theorem 5.2 to generate a sequence of neural networks corresponding to the functions $\mathcal{F}_i$, $1 \leq i \leq N$, and the network for $\mathcal{F}_N$ is the network for the dual of f. In the sequel, the ith application of

Theorem 5.2 refers to the use of Theorem 5.2 to generate the network for
$\mathcal{F}_i$ from the network for $\mathcal{F}_{i-1}$. Therefore, starting from the network for f,
N applications of Theorem 5.2 will yield the network for the dual.

CASE 1:

Consider neuron i. Using Theorem 5.2 to generate networks for func-
tions $\mathcal{F}_1$, $\mathcal{F}_2$, ..., $\mathcal{F}_{i-1}$ increases the threshold of neuron i by $\sum_{j=1}^{i-1} T_{ji}$.
Therefore, in the neural network for function $\mathcal{F}_{i-1}$, the threshold of neuron
i is $I_i + \sum_{j=1}^{i-1} T_{ji}$. After the ith application of Theorem 5.2, the threshold
of neuron i changes to $-(I_i + \sum_{j=1}^{i-1} T_{ji})$ and $-T_{ij}$ is the link-weight of
all neurons $j > i$. The remaining $N - i$ applications of Theorem 5.2 will
increase the threshold of neuron i by $- \sum_{j=i+1}^{N} T_{ji}$. Therefore,

$$I'_i = -(I_i + \sum_{j=1}^{i-1} T_{ji}) - \sum_{j=i+1}^{N} T_{ji}$$

Since $T_{ii} = 0$,

$$I'_i = -I_i - \sum_{j=1}^{N} T_{ji}$$

CASE 2:

Consider two neurons i and j. Without loss of generality, let $j > i$.
The application of Theorem 5.2 to generate networks for functions $\mathcal{F}_1$,
$\mathcal{F}_2$, ..., $\mathcal{F}_{i-1}$ does not change the link weight between neurons i and j.
However, the ith application of Theorem 5.2 changes the weight of the
link between neurons i and j to $-T_{ij}$. The link weight remains $-T_{ij}$ in
networks for all functions $\mathcal{F}_k$, $i \leq k < j$. The jth application of Theorem 5.2 changes the weight of the link between neurons i and j back to T_{ij}.
Note that only the ith and jth applications of Theorem 5.2 can affect the
link weight between neurons i and j. Therefore, starting from the network
for f, after N applications of Theorem 5.2, the weight of the link between
neurons i and j is unchanged, i.e., $T'_{ij} = T_{ij}$.

CASE 3:

The ith application of Theorem 5.2 reduces K by $I_i + \sum_{j=1}^{i-1} T_{ji}$. Therefore, after N applications of Theorem 5.2,

$$\begin{aligned}
K' &= K - \sum_{i=1}^{N}(I_i + \sum_{j=1}^{i-1} T_{ji}) \\
&= K - \sum_{i=1}^{N} I_i - \sum_{i=1}^{N}\sum_{j=1}^{i-1} T_{ji}
\end{aligned}$$

Since $T_{ii} = 0$ and $T_{ij} = T_{ji}$,

$$K' = K - \sum_{i=1}^{N} I_i - \frac{1}{2} \sum_{i=1}^{N} \sum_{j=1}^{N} T_{ji}$$

This completes the proof. ∎

Example 5.5: An OR gate is the dual of an AND gate. Using Theorem 5.3, the neural network $H' = [T'; I'; K']$ for an OR gate can be obtained from the neural network $H = [T; I; K]$ of the AND gate. We proceed as follows. From Table 5.1, with $A = B = 2$, the neural network for the AND gate is

$$T = \begin{pmatrix} 0 & -2 & 4 \\ -2 & 0 & 4 \\ 4 & 4 & 0 \end{pmatrix} \qquad I = \begin{pmatrix} 0 \\ 0 \\ -6 \end{pmatrix} \qquad K = 0$$

From Equation 5.18, $I'_1 = -2$, $I'_2 = -2$ and $I'_3 = -2$. From Equation 5.20, $K' = 0$. From Equation 5.19, it is clear that T' and T are the same. Therefore, H' is given by

$$T' = \begin{pmatrix} 0 & -2 & 4 \\ -2 & 0 & 4 \\ 4 & 4 & 0 \end{pmatrix} \qquad I' = \begin{pmatrix} -2 \\ -2 \\ -2 \end{pmatrix} \qquad K' = 0$$

From Table 5.1, it can easily be verified that T', I' and K' are the weight matrix, threshold vector and constant associated with an OR gate with V_1 and V_2 as inputs, V_3 as the output and $A = B = 2$.

5.4 Three-Valued Model

Recently, Fujiwara has proposed the use of three-valued neurons to model logic circuits [3]. A neuron is viewed as a computing element that can assume one of three states: 0, 1 or 1/2. The first two states are interpreted as before. The third state (1/2) is interpreted as *don't care*, *i.e.*, the signal represented by the neuron can assume either a 0 or 1 value. Traditionally, the value X has been used to indicate that a signal can be either 0 or 1. The purpose of introducing three values is to avoid unnecessary assignment of 0 or 1 values to signals when the signal could assume either value. For example, consider a two-input AND gate with a and b as inputs and c as the output. If we constrain the output to assume the value 0, *i.e.*, $c = 0$, in the binary neuron case, any one of the following three input combinations will set signal c to 0: $(a = 0, b = 0)$, $(a = 0, b = 1)$ or $(a = 1, b = 0)$.

However, in the three-valued neuron case, we only have to choose from the following two possibilities: $(a = 0, b = X)$ or $(a = X, b = 0)$. Therefore, if the AND gate is embedded in a larger circuit, the use of the third value $(1/2)$ reduces the search space for determining a consistent set of signal assignments for the entire circuit.

Fujiwara extends the concepts of energy function and hyperplanes of neurons so that the third value $(1/2)$ is treated sensibly. He also shows that the three-valued model has the same convergence properties as the binary model.

5.5 Summary

We have proposed a novel modeling technique for digital circuits by using neural networks. We will use the neural models and their fundamental properties to:

1. Develop new test generation techniques (Chapters 7, 8, 9 and 10).

2. Identify a new class of easily testable digital circuits (Chapter 11).

3. Identify new instances of NP-complete problems that are easily solvable (Chapters 12 and 13).

References

[1] S. T. Chakradhar, M. L. Bushnell, and V. D. Agrawal. Automatic Test Pattern Generation Using Neural Networks. In *IEEE Proceedings of the International Conference on Computer-Aided Design*, pages 416–419, November 1988.

[2] S. T. Chakradhar, M. L. Bushnell, and V. D. Agrawal. Toward Massively Parallel Automatic Test Generation. *IEEE Transactions on Computer-Aided Design*, 9(8):981–994, September 1990.

[3] H. Fujiwara. Three-valued Neural Networks for Test Generation. In *Proceedings of the 20th IEEE International Symposium on Fault Tolerant Computing*, pages 64–71, June 1990.

[4] J. J. Hopfield. Neurons with Graded Response Have Collective Computational Properties Like Those of Two State Neurons. *Proceedings of the National Academy of Sciences*, 81(10):3088–3092, May 1984.

Chapter 6

TEST GENERATION REFORMULATED

> *"The power of the brain stems not from any single, fixed, universal principle. Instead it comes from the evolution (in both the individual sense and the Darwinian sense) of a variety of ways to develop new mechanisms and to adapt older ones to perform new functions."*

> – M.L. Minsky and S.A. Papert in *Perceptrons*, MIT Press (1988)

We formulate the test generation problem as an optimization problem such that the desired optima are the test vectors for some target fault. This formulation captures three *necessary* and *sufficient* conditions that any set of signal values must satisfy to be a test. First, the set of values must be consistent with each gate's function in the circuit. Second, the signal in the fault-free and faulty circuits at the fault site must assume opposite values (e.g., 0 and 1 respectively, for a s-a-1 fault). Third, for the same primary input vector, the fault-free and faulty circuits should produce different output values.

6.1 ATG Constraint Network

We model the conditions for test generation by joining the circuit-under-test and its *faulty image* (a fault-injected neural network representing the

51

circuit-under-test) back to back at their primary inputs. Primary outputs of the two circuits are connected through an *output interface* to ensure that at least one primary output of the faulty circuit will differ from the corresponding fault-free circuit output. For a single-output circuit, the output interface degenerates into a single inverter whose ports are connected to the primary outputs of the circuit-under-test and the faulty image (see Figure 6.1). It does not matter which way the inverter is connected because it

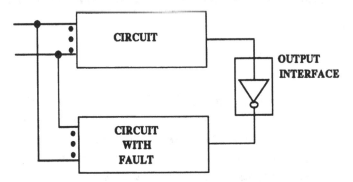

FIGURE 6.1: ATG network for a single-output circuit.

is eventually replaced by a bidirectional neural network. For a circuit with n primary outputs, the output interface is a neural network consisting of n two-input XOR gates and one n-input OR gate. The output of the interface circuit is an OR function of the outputs of the n XOR gates. The inputs to an XOR are the corresponding primary outputs of the circuit-under-test and the faulty image. Figure 6.2 shows the output interface for a circuit with two outputs. The output of the OR gate is assigned the value 1 through-

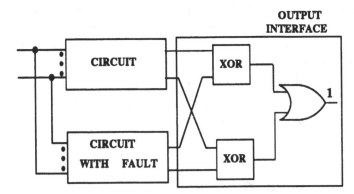

FIGURE 6.2: ATG network for a circuit with two outputs.

out test generation. The circuit-under-test, its faulty image, and the output interface constitute the *ATG constraint network* [3].

6.2 Fault Injection

A fault is injected by modifying the faulty image section of the ATG neural network. Figure 6.3 illustrates the s-a-1 fault in net 3 of the circuit of Figure 5.2. Figure 6.3a is the ATG constraints network and Figure 6.3b, the

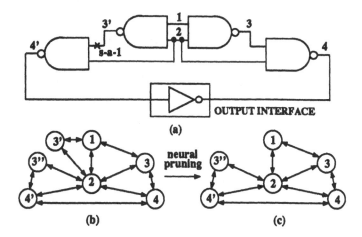

FIGURE 6.3: Fault injected ATG neural networks.

corresponding ATG neural network. Neurons 3′ and 4′ in the faulty image correspond to neurons 3 and 4, respectively, of the circuit-under-test. We have duplicated the neuron representing the net 3′ in the faulty image. The duplicated neuron, 3″, is clamped to the faulty state 1 while neurons labeled 3 and 3′ are fixed to 0. The thresholds of neurons labeled 4 and 4′ and the weight on the link connecting them encode the output interface inverter.

In Figure 6.3c the neural network is simplified by deleting neuron labeled 3′. This is allowed because both neurons 3′ and 3 have the same input values and the same weights and thresholds. This is a simple example of *neural pruning*. A significant reduction in the size of the neural network is possible using the properties of neural network models discussed in Section 5.3. Special considerations are required when a fault is injected in a fan-out net. As an illustration, consider the s-a-1 fault on the input of gate 4 labeled 2 (Figure 6.4a). Figure 6.4b is the corresponding ATG neural network with the fault injected. We have duplicated the neuron representing the primary input net 2. The duplicated neuron, 2′, is clamped to the faulty state 1 while the neuron labeled 2 is fixed at 0. Therefore, the input net 2 of gate 4 will have a logic value 0 in the fault-free circuit and a logic value 1 in the faulty circuit. Neural pruning collapses the neurons labeled 3 and 3′ since they always have the same logic value. The simplified network is shown in Figure 6.4c.

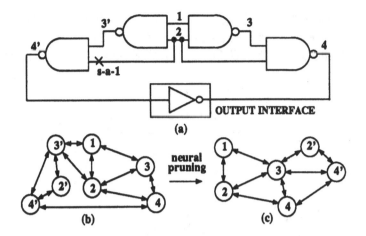

FIGURE 6.4: Fault injection in a fanout stem.

6.3   Test Generation

When a fault is injected into the ATG neural network the output interface incorporates the constraint that at least one primary output of the faulty circuit will differ from the corresponding fault-free circuit output. Therefore, if the fault is testable, there exists a *consistent labeling* of the neurons in the ATG neural network with values from the set {0,1} that does not violate the functionality of any gate. To obtain a test vector for a given fault, we have to find a minimum energy state of the ATG neural network. The activation values of the primary input neurons in the minimum energy state of the ATG network form the test vector for the fault.

6.4   Summary

Having discussed the basic formulation of test generation as an energy minimization problem, we will develop several new test generation techniques. Methods of test generation using the energy minimization formulation are based upon:

1. Use of an actual neural network [1, 4].

2. Use of single or multiple processor computers to simulate the neural network.

Large scale implementations of neural networks are presently not available and, therefore, we will mostly rely on simulation. We explore gradient-descent and probabilistic relaxation methods for test generation in Chapter 7. A continuous optimization technique for test generation is investigated in Chapter 8 by simulating an analog neural network on a commercial neurocomputer [2]. We present graph-theoretic techniques for test generation in Chapters 9 and 10. Enhancements to the basic test generation formulation vastly improve the efficiency of the proposed test generation techniques.

References

[1] J. Alspector and R. B. Allen. A Neuromorphic VLSI Learning System. In P. Loseleben, editor, *Advanced Research in VLSI: Proceedings of the 1987 Stanford Conference*, pages 313–349. MIT Press, Cambridge, Massachussetts, 1987.

[2] ANZA – Neurocomputing Coprocessor System. *Computer*, 21(9):99, September 1988.

[3] S. T. Chakradhar, M. L. Bushnell, and V. D. Agrawal. Toward Massively Parallel Automatic Test Generation. *IEEE Transactions on Computer-Aided Design*, 9(8):981–994, September 1990.

[4] C. D. Kornfeld, R. C. Frye, C. C. Wong, and E. A. Rietman. An Optically Programmed Neural Network. In *Proceedings of the International Conference on Neural Networks, San Diego, CA*, volume 2, pages 357–364, July 1988.

Chapter 7

SIMULATED NEURAL NETWORKS

"Some day, the sun will grow cold, and life on the earth will cease. The whole epoch of animals and plants is only an interlude between ages that were too hot and ages that will be too cold. There is no law of cosmic progress, but only an oscillation upward and downward, with a slow trend downward on the balance owing to to the diffusion of energy."

– B. Russell in *Religion and Science*, Oxford University Press (1974)

Large scale implementations of neural networks are presently not available. However, serial and parallel computers have been used to simulate neural networks [2, 5]. We describe algorithms for simulating the neural network on a serial computer and discuss possible implementations of these algorithms on parallel computers. Chapter 8 discusses test generation on a commercial hardware accelerator for neural network simulation.

7.1 Iterative Relaxation

For the test generation problem, we attempt to find the global minimum of the energy function by using a fast gradient descent search algorithm. In our formulation, the energy should be 0 at all global minima. If the search terminates at a local minimum, i.e., the final energy E of the network is non-zero, we use a relaxation technique to determine the global minimum.

We first describe the gradient descent technique. The difference between the energy of the neural network with the kth neuron *off* ($V_k = 0$) and its energy with that neuron *on* ($V_k = 1$) is derived from Equation 5.1 as follows:

$$\begin{aligned} E_k &= E(V_k = 0) - E(V_k = 1) \\ &= I_k + \sum_{j=1}^{N} T_{jk} V_j \end{aligned}$$

where E_k is known as the *net input* to neuron k. It is apparent that the effect of changing the state of neuron k on the total energy E is completely determined by the states of all neurons connected to neuron k and by the corresponding link weights. Consequently, the differences E_k for various k can be computed locally and in parallel. The updating rule is to switch a neuron into whichever of its two states yields the lower total energy of the neural network, given the current states of all other neurons. Each neuron may update stochastically at some average rate using the update rule:

$$V_k = \begin{cases} 1 & \text{if } E_k > 0 \\ 0 & \text{if } E_k < 0 \\ V_k & \text{otherwise.} \end{cases} \tag{7.1}$$

If $E_k > 0$ ($E_k < 0$), the neural network state with $V_k = 1$ ($V_k = 0$) has a lower energy than the state with $V_k = 0$ ($V_k = 1$). Since the objective is to minimize the energy of the neural network, neuron k will update to assume the activation value $V_k = 1$ ($V_k = 0$). If $E_k = 0$, the neural network states with $V_k = 1$ and $V_k = 0$ have the same energy and neuron k retains its old activation value. Since $T_{ij} = T_{ji}$ and $T_{ii} = 0$ the discrete stochastic algorithm always converges to a (local or global) minimum of E [7]. The state space of the network consists of 2^N corners of the N-dimensional hypercube.

We illustrate the gradient descent technique by an example. Note that the energy of the neural network never increases during energy minimization even though the number of inconsistently labeled gates may increase. Consider the circuit shown in Figure 7.1a. Without loss of generality, let $A = B = 2$ be the constants for all gates in the circuit. The corresponding neural network is shown in Figure 7.1b. From Table 5.1, the gate energy functions can be computed and the energy function for the circuit is:

$$\begin{aligned} E_{CKT} &= E_{OR}(V_p, V_q, V_r) + E_{NAND}(V_r, V_q, V_s) \\ &\quad + E_{OR}(V_r, V_s, V_t) + E_{OR}(V_s, V_q, V_u) \end{aligned}$$

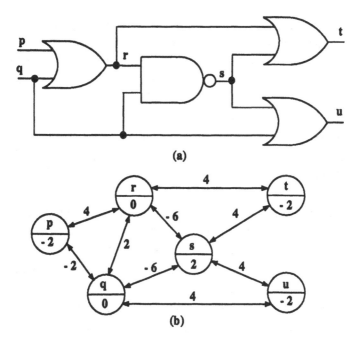

(a)

(b)

FIGURE 7.1: An example circuit (a) and the corresponding neural network (b).

Initially, let all neurons have an activation value of 0. Thus, $E_{CKT} = 6$. Recall that neuron activation values correspond to logic values of signals. Only NAND gate s is inconsistently labeled. We commence the gradient descent phase by randomly selecting a neuron for updating. If any of the neurons p, q, r, t or u is selected for updating, the state of the network does not change. This is because the net input $E_k = E_{CKT}(V_k = 0) - E_{CKT}(V_k = 1)$ of each of these neurons is less than or equal to 0 and all neurons already have an activation value 0. In other words, changing the activation value of any one of the neurons p, q, r, t or u does not decrease the neural network energy. When neuron s (see Figure 7.1b) is selected for updating, the state of the network changes. The net input E_s to the neuron s is 2 and the neuron is updated to $V_s = 1$. The energy of the network is reduced to $E_{CKT} = 4$. Note that although the number of inconsistent gates has increased (OR gates with outputs t and u now have inconsistent port values), the total energy of the network has been reduced. At this stage, if any of the neurons p, q, r or s are selected for updating, the state of the network remains unchanged. Suppose neuron t is selected for update. Since the net input E_t is 2, V_t is updated to 1 and the energy of the network reduces to $E_{CKT} = 2$. When neuron u updates, $V_u = 1$, $E_{CKT} = 0$ and a consistent labeling is achieved.

Indeed, the gradient descent algorithm can also terminate at a local minimum due to the fact that it is a greedy algorithm and only accepts moves which reduce the energy of the neural network. To avoid this behavior, probabilistic algorithms can be devised to also accept some moves that increase the neural network energy. The criterion for accepting such moves is generally based on the increase in the energy they cause. Since the final objective still is to minimize energy, the higher-energy moves are only accepted in a probabilistic sense. Thus, higher-energy moves are allowed but lower-energy moves are statistically favored. Probabilistic search algorithms are also referred to as simulated annealing methods [11, 18].

Simulated annealing is analogous to the crystallization process. A crystal is a minimum-energy configuration of molecules. We melt the substance and then slowly lower its temperature until the crystal forms. The rate of temperature decrease must be very slow as we approach solidification.

The Monte Carlo method [14] has been used to simulate the annealing process and has also been proposed as an effective probabilistic method for finding a global minimum in combinatorial optimization problems. In the latter case, we randomly sample the state-space and accept a sample depending upon whether the energy of the sample satisfies some acceptance criterion based on a controlling parameter T. Here, T is analogous to temperature in the annealing process. In this section we describe a sequential implementation of the probabilistic method and in Section 7.3 we show how this procedure can be extended to parallel implementations.

Markov chains (see [1, 18, 10] for a detailed description) have been used as a mathematical model for describing how a probabilistic method finds a global minimum in optimization problems. A Markov chain consists of a sequence of states where the next state probabilistically depends only on the current state. The mechanics of the probabilistic method can be summarized as follows. Starting from a chosen initial state and an initial value of T, a sequence of states is generated. These states form the first Markov chain. The probabilistic method lowers the value of T and generates a new Markov chain whose initial state is the final state of the previous chain. Similarly, the value of T is successively lowered and the corresponding Markov chains are generated. This continues until T has assumed a sufficiently low value. It can be shown that under certain restrictions on the cooling process, the probabilistic method asymptotically produces an optimal solution to the combinatorial optimization problem with probability 1 [1].

For the test generation problem, we use the probabilistic method to find a global minimum of the energy function whenever the gradient descent method terminates at a local minimum. Given a state S_1 of the neural network in which the kth neuron has activation value V_k, a neighboring state S_2 is obtained by changing the activation value of the kth neuron to $1 - V_k$. Without loss of generality, let S_1 (S_2) be the state of the neural network in which the kth neuron has activation value 1 (0). The energy difference between the two states is computed as $E_k = E(S_2) - E(S_1)$. The *acceptance probability* p_k is the probability that the state with $V_k = 1$ is preferred over the state with $V_k = 0$, and is given by:

$$p_k \;=\; \frac{1}{1 + e^{-E_k/T}}$$

A random number r between 0 and 1 is generated and if $r \leq p_k$, state S_1 is accepted as the next state. Otherwise, state S_2 is accepted.

For minimization of the energy function E by the probabilistic method we start at an initial value of the parameter T with some (possibly random) initial state of the neural network. A sequence of states (constituting a single Markov chain) is generated at the initial value of the parameter T. The value of T is lowered and a new Markov chain is generated. T is successively lowered to generate subsequent Markov chains. Eventually as T approaches 0, state transitions become more and more infrequent and, finally, the neural network stabilizes in a state which is taken as the final solution. In practical implementations of the algorithm, however, the asymptoticity conditions may never be attained and thus convergence to an optimal solution is not guaranteed. Occasionally, the algorithm obtains a locally optimum solution that may be close to the optimal solution instead of being the optimal solution. Consequently, it is an approximation algorithm. The initial value of T, the number of states in the Markov chain at each value of T and the rate of decrease of T are all important parameters that affect the speed of the algorithm and the quality of the final solution.

Unlike many classical optimization problems, such as the traveling salesperson problem [13], the optimum value of the energy function E for the test generation problem is known *a priori*. This knowledge can be a powerful aid in the analysis of the probabilistic algorithm [1]. Furthermore, this information provides an effective criterion to stop the probabilistic algorithm. If, during the relaxation process, the network energy E ever assumes its minimum value 0, we can stop the relaxation process since this is the global minimum. This is in contrast to the criterion typically used in terminating the relaxation in many other classical optimization problems where one stops the relaxation process only when the temperature has assumed a sufficiently low value. This is because, in general, the minimum value of energy for the problem may not be known beforehand and hence, it is not known how close one is to the optimal solution.

7.2 Implementation and Results

Our test generation scheme can be implemented directly on an actual neural network. Programmable neural networks have been developed using VLSI [3] and optical technology [12]. However, large scale neural networks are, as yet, not a reality. Alternatively, single-processor and multi-processor computers are used to simulate neural networks [8]. The results presented in this section are based on a single-processor computer simulation of the neural network. In Section 7.3, we discuss the simulation of the neural network on a parallel computer.

7.2.1 Test Generation System

Figure 7.2 shows our test generation system. The *neural compiler*, using the *neural database*, maps the circuit onto the ATG neural network discussed in Chapter 6. The *fault specific transformation stage* introduces a modeled fault in the *fault database*, performs fault-specific modifications, and eliminates some neurons (*neural pruning*) from the ATG neural network. Although not used in the present system, circuit partitioning techniques can further reduce the size of the neural network. For example, only devices that lie on paths from the fault site to any of the primary outputs may be explicitly included in the faulty image. The basic idea of pruning is to reduce the number of neurons in the ATG neural network and speed up the relaxation process. Initial conditions for relaxation are obtained from logic simulation. The *iterative relaxation stage* relaxes the neural network to generate test vectors.

7.2.2 Experimental Results

To demonstrate the feasibility of our test generation technique, we generated tests for several combinational circuits. Three of them, the Schneider's circuit [16], the ECAT (error correction and translation) circuit [6] and the SN5483A [17] four-bit carry look-ahead binary full adder, are included here. Our experimental ATG system does not have a fault simulator. Also, no fault collapsing or circuit partitioning techniques are used. Tests were generated independently for every single stuck-at fault at the inputs and output of each gate. A preliminary version of the ATG system was implemented in C to run on a SUN 3/50 workstation.

Results in this section refer to values of $A = B = J = 200$ and $C = 100$ for the constants in the models for the logic gates (see Table 5.1). The exact values of these constants are not critical for the gradient descent

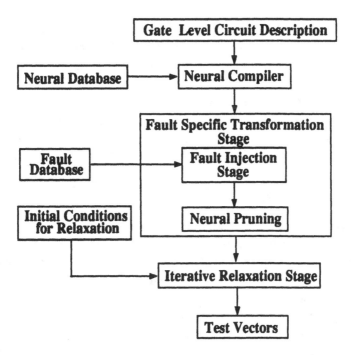

FIGURE 7.2: The ATG system.

method because the neurons change states based on the sign of the energy differences rather than on the magnitude of the energy differences. However, for the probabilistic relaxation technique, care is required in choosing the constants as they influence the initial value of the parameter T.

The initial value T_0 of the parameter T must be large to ensure access to all possible states of the neural network with high probability. This requirement is necessary (but not sufficient) to prevent the minimization process from terminating at a local minimum. If T_0 is too small in comparison to E_k values (corresponding to state changes that allow the neural network to escape from the local minimum), these state changes will probably not be accepted and consequently the energy minimization process will get trapped in some local minimum. To our knowledge, the published literature does not provide a basic formalism to uniquely determine the initial value T_0. An adequate heuristic expression for T_0 is given by $T_0 = 2 \times g \times (A + B + 2C)$ where g is the number of gates in the circuit. The expression $2 \times (A + B + 2C)$ is an upper bound on the maximum energy of a gate in the basis set. This starting value clearly guarantees access to all possible states with sufficiently high probability. Other heuristics have also been proposed for determining the initial value T_0 [1].

The decrement of the temperature parameter T determines the behavior of the minimization process in the limit $T \rightarrow 0$. Again, the literature, to our knowledge, does not provide a basic formalism from which a unique value for the decrement can be determined. Experimental observations reveal that the decrement of the parameter T is not very critical as long as the Markov chains are long, i.e., the number of neuron state changes attempted at a given value of T is large. For N neurons in the network and T_0 as the initial value of the parameter T, we attempted N neuron state changes when $\frac{2T_0}{3} \leq T \leq T_0$, $2N$ neuron state changes when $\frac{T_0}{3} \leq T \leq \frac{2T_0}{3}$ and $4N$ state changes when $0 \leq T \leq \frac{T_0}{3}$. An annealing schedule, $T_n = \alpha^n T_0$, with $\alpha = 0.95$ was used.

As stated earlier, initial conditions for the gradient descent and the probabilistic relaxation phase were obtained from logic simulation. Starting from an arbitrary vector of values for the primary inputs of the circuit, all neurons in the network were initialized with signal values obtained from logic simulation. A fault was injected and gradient descent search was commenced. If the search was successful, the consistent labeling corresponding to the test vector was used as the initial condition for the gradient descent search for the next fault. Gradient descent search was thus conducted once for each fault. For the faults not detected due to local minima encountered in the gradient descent method, we used the probabilistic relaxation method. For practical reasons an upper limit was set on the permitted number of neuron state changes during relaxation. Once the relaxation method found a test vector, using the corresponding initial conditions, we again used the gradient descent method and relaxation, if necessary, to cover the remaining undetected faults. Occasionally, a fault may remain undetected due to the preset state change limit in the relaxation method. This combination of the two methods was repeatedly applied until each fault was either detected or classified as undetectable due to the limitations of the system.

We generated tests for all faults in the Schneider's circuit[1] and the four-bit parallel adder (SN5483A). In the ECAT circuit, there were eight redundant faults which were the only faults left undetected after the limit of 5,000 neuron updates was reached. Table 7.1 summarizes the results. Note that even though we did not use any fault simulator, the number of tests is smaller than the number of faults. This is because several of the generated tests were identical.

---

[1]This is the original Schneider's circuit in which all faults are detectable. In later chapters we use a modified version with four redundant faults.

TABLE 7.1: Experimental results for the gradient descent method augmented by relaxation.

Circuit	Schneider	ECAT	SN5483A
Number of Signals	12	15	49
Number of Inputs	4	6	9
Number of Outputs	3	1	5
Number of 2-input gates	11	26	71
Tests Generated	9	9	36
Total Faults	24	30	98
Faults Detected	24	22	98
Total Testable Faults	24	22	98
% of All Testable Faults Detected	100	100	100
Average Time per Fault (sec.)	0.12	0.43	3.3

7.3 Parallel Simulation

Large scale implementations of neural networks are presently not available. However, parallel computers have been used to simulate neural networks [2, 4, 5]. As mentioned in Section 7.1, neurons only require local information to calculate their state transitions. Thus, many neurons can change state in parallel. We must, however, distinguish between *synchronous* and *asynchronous parallelism*. In the following discussion, we summarize these two techniques used for neural network simulation.

7.3.1 Synchronous Parallelism

In the synchronous model, neurons compute the energy differences E_k and change their states at discrete periodic intervals. In both the gradient descent and the probabilistic method, the kth neuron has to perform two sequential tasks:

1. Calculation of the energy difference E_k.

2. Transition to a new state.

Since the kth neuron only needs to know the states of its neighboring neurons and the corresponding link weights to calculate the energy difference E_k, *all* neurons can concurrently compute their energy differences.

It is desirable that *all* neurons be capable of changing their states in parallel as well. The probabilistic method will converge to a global minimum even when *all* neurons simultaneously change their states [2]. The convergence in the gradient descent method, however, needs closer examination.

Gradient Descent: In the serial implementation of the gradient descent method, a neuron state transition guarantees a decrease in the energy of the network and the method generates a sequence of neural network states with decreasing energy. It is important that the convergence to minimum energy of the serial gradient descent method be retained in the parallel version. Given a state S_1 of the neural network, if *all* neurons have their state transitions in parallel, then the new state S_2 may have a *higher* energy than the initial state S_1 and the gradient descent method may not converge (i.e., although each neuron assumes a state that will decrease the energy of the network, simultaneous state changes of multiple neurons may not result in a net reduction of the network energy). As an illustration, consider the neural network for the AND gate shown in Figure 5.1a. Initially, let the activation values of neurons x_1, x_2 and x_3 be 1, 1 and 0, respectively. The energy of the network is $E = B = 2$. All three neurons compute their respective energy differences $E_{x_1} = -B = -2$, $E_{x_2} = -B = -2$ and $E_{x_3} = 2(A + B) - 2A - B = 2$, in parallel. If neurons x_1 and x_2 simultaneously change states (neuron x_3 does not change state), then neurons x_1, x_2 and x_3 will assume activation values 0, 0 and 1, respectively, resulting in an increase of network energy to $E = 2A + B = 6$. Therefore, all three neurons in the AND gate cannot simultaneously change states and still guarantee a decrease in the network energy. In fact, only the neurons that are not connected to each other should be allowed to change their states in parallel.

For the gradient descent method, synchronous parallelism can be obtained by partitioning the set of neurons into disjoint subsets such that neurons in the same subset are not connected to each other. A clocking scheme is introduced and in each clock cycle, the neurons in a randomly chosen subset are allowed to simultaneously change their states. For optimal exploitation of parallelism, the number of subsets should be as small as possible. The problem of finding a minimal partition is known as the *graph coloring* problem [9]. The amount of parallelism that can be obtained strongly depends on the connectivity among neurons and will be relatively small if the connectivity is dense. Furthermore, this mode of parallelism requires some form of global synchronization among neurons because there will be connections between neurons in different subsets.

Probabilistic Relaxation: In this method, *all* neurons can change their states in parallel at discrete periodic intervals but the method still ensures convergence to the global minimum of the energy function. It can be shown [1] that the probability of a neuron changing its state decreases as the value of the parameter T decreases, and this probability approaches 0 as T approaches 0. Thus, for smaller values of T, the probability that two or more neighboring neurons simultaneously change their states becomes negligibly small. Also, an occasional energy increase is acceptable in the probabilistic framework.

7.3.2 Asynchronous Parallelism

Asynchronous parallelism is based on the assumption that a neuron can change its state at any time, and thus, at random intervals. It would be difficult to exploit asynchronous parallelism for the gradient descent method. As we have seen in Section 7.3.1, simultaneous updates of some neurons may cause an increase in the energy and, therefore, global synchronization is necessary to ensure convergence to minimum energy. The probabilistic method can, however, exploit such parallelism and the convergence of a parallel implementation to global energy minimum is guaranteed. The proof is similar to the synchronous parallelism case described in Section 7.3.1. A distinct advantage of asynchronous parallelism is that a global synchronization is no longer necessary.

7.4 Summary

A new algorithm for generating test vectors based on Hopfield neural networks and relaxation techniques is feasible. Due to the inherent properties of such networks, the algorithm is suitable for massively parallel execution. We have formulated the test generation problem as an optimization problem and a whole suite of techniques available for the well-studied quadratic 0-1 optimization problem can be used for a solution. Preliminary results using serial computer simulation of neural networks have established the feasibility of this method. Since Hopfield networks are bidirectional, the algorithm can potentially be extended to handle bidirectional fault propagation in switch-level and mixed-level descriptions of VLSI circuits [15]. With advances in technology, when large-scale neural networks become a reality, our technique should provide a significant advantage over other methods.

References

[1] E. H. Aarts and P. J. van Laarhoven. Statistical Cooling: A General Approach to Combinatorial Optimization Problems. *Philips Journal of Research*, 40(4):193–226, September 1985.

[2] E. H. L. Aarts and J. H. Korst. *Simulated Annealing and Boltzmann Machines: A Stochastic Approach to Combinatorial Optimization and Neural Computing*. Wiley Interscience Series in Discrete Mathematics, John Wiley and Sons, New York, 1989.

[3] J. Alspector and R. B. Allen. A Neuromorphic VLSI Learning System. In P. Loseleben, editor, *Advanced Research in VLSI: Proceedings of the 1987 Stanford Conference*, pages 313–349. MIT Press, Cambridge, Massachussetts, 1987.

[4] S. E. Fahlman, G. E. Hinton, and T. J. Sejnowski. Massively Parallel Architectures for AI: NETL, Thistle and Boltzmann Machines. In *Proceedings of the National Conference on AI, AAAI-83*, pages 109–113, August 1983.

[5] J. Ghosh and K. Hwang. Mapping Neural Networks onto Highly Parallel Multiprocessors. Technical Report CRI-87-65, University of Southern California, Computer Research Institute, Los Angeles, 1987.

[6] P. Goel. An Implicit Enumeration Algorithm to Generate Tests for Combinational Logic Circuits. *IEEE Transactions on Computers*, C-30(3):215–222, March 1981.

[7] J. J. Hopfield. Neurons with Graded Response Have Collective Computational Properties Like Those of Two State Neurons. *Proceedings of the National Academy of Sciences*, 81(10):3088–3092, May 1984.

[8] J. J. Hopfield and D. Tank. Neural Computation of Decisions in Optimization Problems. *Biological Cybernetics*, 52(3):141–152, July 1985.

[9] E. Horowitz and S. Sahni. *Fundamentals of Computer Algorithms*. Computer Science Press, Rockville, Maryland, 1984.

[10] S. Karlin. *A First Course in Stochastic Processes*. Academic Press, New York, 1973.

[11] S. Kirkpatrick, C. D. Gelatt, and M. P. Vecchi. Optimization by Simulated Annealing. *Science*, 220(4598):671–680, May 1983.

[12] C. D. Kornfeld, R. C. Frye, C. C. Wong, and E. A. Rietman. An Optically Programmed Neural Network. In *Proceedings of the International Conference on Neural Networks, San Diego, CA*, volume 2, pages 357–364, July 1988.

[13] S. Lin and B. W. Kernighan. An Efficient Heuristic for the Traveling Salesman Problem. *Operations Research*, 21(2):498–516, March-April 1973.

[14] N. Metropolis, A. W. Rosenbluth, M. N. Rosenbluth, and A. H. Teller. Equation of State Calculations by Fast Computing Machines. *Journal of Chemical Physics*, 21(6):1087–1093, 1953.

[15] M. K. Reddy, S. M. Reddy, and P. Agrawal. Transistor Level Test Generation for MOS Circuits. In *Proceedings of the 22nd ACM/IEEE Design Automation Conference*, pages 825–828, 1985.

[16] P. R. Schneider. On the Necessity to Examine D-Chains in Diagnostic Test Generation. *IBM Journal of Research and Development*, 11(1):114, January 1967.

[17] *The TTL Data Book for Design Engineers, Second edition*, page 199. Texas Instruments, 1973.

[18] P. J. M. van Laarhoven and E. H. L. Aarts. *Simulated Annealing: Theory and Applications*. Kluwer Academic Publishers, Dordrecht, The Netherlands, 1987.

Chapter 8

NEURAL COMPUTERS

"The original question, 'Can machines think?', I believe to be too meaningless to deserve discussion. Nevertheless I believe that at the end of the century the use of words and general educated opinion will have altered so much that one will be able to speak of machines thinking without expecting to be contradicted."

– A.M. Turing in "Computing Machinery and Intelligence", *Mind*, Vol. LIX, No. 236 (1950)

In the previous chapters, we developed our theory assuming a digital model of a neuron. This is useful as it allowed us to solve the test generation problem for digital circuits. In later chapters, we will research several applications of the discrete models. However, the real neurons are analog elements and we must examine how well the present solution will work if actual neural network hardware was available. In this chapter, we present a solution to the test generation problem using a neurocomputer that contains special hardware to perform energy minimization for analog neural networks [4]. We also wrote a computer program on a SUN 3/50 workstation to simulate the network. Both results are in agreement and demonstrate the feasibility of our approach.

We assume that a signal can assume any value between 0 and 1 during test generation. This is in contrast to the the classical test generation

71

algorithms [1] where signals only assume discrete values. This is the first application of neural computer hardware to the test generation problem. Due to the limited capacity of the present-day neurocomputers, only small circuits could be processed by the hardware.

8.1 Feasibility and Performance

Once the test generation problem is programmed on a neural network, two questions arise. The first is that of *feasibility*. Since the solution requires an absolute energy minimization, feasibility here means how well our formulation can avoid local minima of the energy function. We answer that question in this chapter through hardware and software simulations.

The second question is that of *performance*, that is, how fast is the test generation. This cannot be answered easily. The settling time of a real neural network depends on the time constants of the analog computing elements comprising the network. When the problem size (circuit size) increases, one adds more neurons to solve it. The connectivity of an individual gate in a digital circuit grows more slowly than the circuit size. Therefore, the time constant of the corresponding neural network will also increase at a slower rate than the circuit size. The settling time of fairly large neural networks has been reported as 0.1 nanosecond [12]. Unfortunately, in that work, the authors do not explain their estimation technique. As technology advances, more data will become available and it is clear that neural networks will provide massively parallel solutions to this and many other problems.

8.2 ANZA Neurocomputer

Programmable neural networks have been developed using VLSI [3] and optical technologies [14]. However, large scale neural networks are, as yet, not a reality and, therefore, most of the work on neural networks relies on digital simulation using general-purpose serial or parallel computers, or special-purpose processors. A *Neurocomputer* contains special-purpose hardware for high-speed neural network simulation. Typically, a combination of general-purpose microprocessors and digital signal processing integrated circuits is used in the design of neurocomputers [5].

A neurocomputer usually consists of one or more digital processing elements. A fully-interconnected neurocomputer has a physical processing element for every neuron in the network and each processing element is directly connected to every other processing element through an interconnection network. Such systems have two disadvantages: (1) the interconnection network becomes prohibitively large, and (2) the processing elements are not efficiently used. As a result, most commercial neurocomputers are not fully-interconnected. They have a small number of physical processing elements, with a subset of neurons being processed by each processing element. This architecture provides an efficient use of the available processing resources. A basic property of the analog neural networks is that the minimization of the energy function E (Equation 4.7) is equivalent to solving the system of differential equations given by Equation 4.6. Thus, a fundamental requirement of a neurocomputer is its ability to solve a large set of coupled differential equations [17]. Since most existing neurocomputers have digital processors, these differential equations are solved by numerical techniques. Alternative implementations may use energy minimization methods like Newton-Raphson or other fixed-point methods [11].

The ANZA is a commercially available neurocomputing co-processor board from Hecht-Nielsen Neurocomputers. Different versions can be installed in a personal computer (e.g., an IBM PC/AT) or a workstation (e.g., a SUN 3/50) host. Operating in parallel with the host computer, the ANZA board off-loads all neural network processing from the host. The co-processor board has a four-stage pipelined Harvard architecture with separate data and instruction paths. It consists of a Motorola MC68020 microprocessor with access to four megabytes of dynamic RAM. To improve the floating point performance, a Motorola MC68881 floating point co-processor is used. Both processors operate at 20 MHz. Interaction between user software and the ANZA is accomplished through a set of callable subroutines.

The ANZA solves a system of N differential equations, for given initial values, by using the Euler method [11]. For this purpose, the differential equation given by Equation 4.6 is replaced by a difference equation suitable for iterative processing. The $(k + 1)$th iteration is described as follows:

$$u_i^{(k+1)} = u_i^{(k)} + h \times f(u_1^{(k)}, \ldots, u_N^{(k)}), \ 1 \leq i \leq N \qquad (8.1)$$

where h is a small time step and

$$f(u_1, \ldots, u_N) = \sum_{j=1}^{N} T_{ij} V_j - \frac{u_i}{\tau} + I_i. \qquad (8.2)$$

Starting from some initial values $u_i^{(0)}$, the value of u_i is iteratively updated. The time constant τ determines the convergence time of an actual neural network. Since the convergence time is in arbitrary units, we can assume $\tau = 1$ [10]. Also, h is the step size in the iterative solution. The

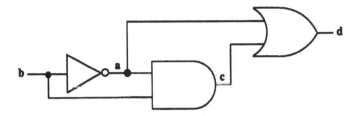

FIGURE 8.1: An example circuit.

following transfer function is often assumed [10]:

$$V_i = \frac{1}{2}(1 + \tanh(\lambda u_i)) \qquad (8.3)$$

To solve an optimization problem on the ANZA, we determine the link weights and the thresholds from the the energy function (Equation 4.2) and solve the system of difference equations described by Equation 8.1 using the software provided by ANZA.

8.3 Energy Minimization

We illustrate the new minimization technique by an example. Consider the problem of generating a test for the s-a-1 fault on signal d in the circuit of Figure 8.1. Since the fault is on a primary output, we need not create a faulty circuit copy. We constrain d to assume the value 0. Thus, the gate energy functions [7] are:

$$\begin{aligned}
E_{NOT}(V_b, V_a) &= 4V_aV_b - 2V_a - 2V_b + 2 \\
E_{AND}(V_a, V_b, V_c) &= -4V_aV_c - 4V_bV_c + 2V_aV_b + 6V_c \\
E_{OR}(V_a, V_c, V_d) &= -4V_aV_d - 4V_cV_d + 2V_aV_c + 2V_a + 2V_c + 2V_d
\end{aligned}$$

The energy function for the circuit is:

$$E_{CKT} = E_{NOT}(V_b, V_a) + E_{AND}(V_a, V_b, V_c) + E_{OR}(V_a, V_c, V_d)$$

We can rewrite E_{CKT} so that it has the same form as Equation 4.2 and the corresponding neural network is shown in Figure 8.2. Since signal d is constrained to be 0, we only have to find the values of V_a, V_b and V_c that will minimize E_{CKT}. There are three unknown variables in E_{CKT} and an analog neural network of three neurons is required to minimize the energy function. An iterative minimization of E_{CKT} approximates the time evolution of the three-neuron analog neural network described by the following

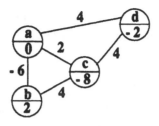

FIGURE 8.2: Neural network corresponding to circuit in Figure 8.1.

three first order differential equations (derived from Equation 4.6):

$$\frac{du_a}{dt} = [-6V_b + 2V_c + 4V_d] - u_a$$

$$\frac{du_b}{dt} = [-6V_a + 4V_c] - u_b + 2$$

$$\frac{du_c}{dt} = [2V_a + 4V_b + 4V_d] - u_c - 8$$

V_a, V_b and V_c are the output voltages and u_a, u_b and u_c are the input voltages of neurons a, b and c, respectively. As stated earlier, the transfer function of a neuron is given by Equation 8.3. The first order system can be approximated by the following difference equations:

$$u_a = u_a + h \times (-6V_b + 2V_c + 4V_d - u_a)$$
$$u_b = u_b + h \times (-6V_a + 4V_c - u_b + 2)$$
$$u_c = u_c + h \times (2V_a + 4V_b + 4V_d - u_c - 8)$$

Starting with some initial values for V_a, V_b and V_c, and some suitable values for λ and h, the method of Section 8.2 can be employed to solve these equations. Note that every differential equation in the system is discretized identically and, therefore, the step size h must be sufficiently small so that the fastest changing variable in the system is accurately represented. For example, starting from the initial values $V_a = V_b = V_c = 0.5$, $\lambda = 50$ and $h = 0.0001$, the Euler method yields the solution $V_a = V_c = 0$, $V_b = 1$. This is a test for the fault because at these values E_{CKT} evaluates to 0. For some other initial states or some large step size, the method may yield solutions that are local minima of the energy function and, consequently, we may not find a test for the fault.

8.4 Enhanced Formulation

Since the time evolution of the neural network is a greedy minimization of the energy function (Equation 4.2), the neural network can sometimes settle in a stable state corresponding to a local minimum. In such cases, the energy function does not evaluate to 0 and, consequently, no test is found for the fault under consideration. Local minima can be avoided to a large extent by adding problem-specific knowledge to the energy function derived from (1) circuit topology and (2) the fact that a test sensitizes a path from the fault site to a primary output of the circuit. We discuss two techniques to enhance the energy function. Both techniques are easily parallelizable.

8.4.1 Transitive Closure

In Figure 8.1, it is easy to see that if $V_b = 1$, then V_d assumes the value 0, and if $V_b = 0$, then $V_d = 1$. These two constraints can be incorporated into E_{CKT} by adding the terms $2V_bV_d$ and $2(1 - V_b)(1 - V_d)$. If $V_b = 1$, then the term $2V_bV_d$ will be zero only for $V_d = 0$. Also, if $V_b = 0$, then the term $2(1 - V_b)(1 - V_d)$ will be zero only for $V_d = 1$. Note that signal c can be fixed at 0, i.e., V_c assumes the value 0 for all input vectors. Identifying such signals reduces the number of variables in the energy function and, hence, decreases the search space of the minimization problem. As we will now explain, all fixed signal values and all pairwise relationships between signals in the circuit can be derived in a systematic manner as follows.

To establish relationships among signals, consider the AND gate shown in Figure 8.3. Its energy function can be written as:

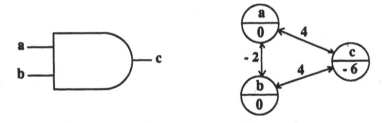

FIGURE 8.3: AND gate (a) and its neural net model (b).

$$E_{AND}(V_a, V_b, V_c) = [2V_c(1 - V_a)] + [2V_c(1 - V_b)]$$
$$+ [2V_c - 2V_aV_c - 2V_bV_c - 2V_aV_b]$$

The three parts shown in brackets will be referred to as terms of the energy function. It can easily be verified that each term independently becomes 0 at all values of V_a, V_b and V_c that conform to the AND gate truth table. The first term is 0 only when V_a and V_c satisfy the relation $V_c \Rightarrow V_a$ (i.e., V_c implies V_a). Similarly, the second term is 0 only when $V_c \Rightarrow V_b$. The third term cannot be represented as a binary relation. Therefore, two binary relations can be derived for the AND gate. Similar relations are possible for other types of gates. For a circuit with g two-input gates, we have a set R consisting of $2 \times g$ binary relations that are satisfied by any assignment of signal values corresponding to a test. R may imply the following types of logical conclusions:

- *Contradiction:* When no assignment of signal values can simultaneously satisfy all relations in R, the fault is *redundant*.

- *Fixation:* Some signal x can be fixed to the value 0 or 1, i.e., the signal assumes only one logic value in all tests for the fault under consideration.

- Non-adjacent pairwise signal relationships.

All contradictions, fixations and non-adjacent pairwise signal relationships can be determined by finding the *transitive closure* of R [2]. The set R^T, the transitive closure of R, consists of all relations in R, and if two binary relations $x \Rightarrow y$, $y \Rightarrow z$ (x, y and z are Boolean variables) belong to set R, then the relation $x \Rightarrow z$ is included in R^T. Contradictions and fixations are now determined from R^T [9]. If R^T consists of the relations $x \Rightarrow \overline{x}$ and $\overline{x} \Rightarrow x$ for some variable x, then we have a contradiction since no set of signal values can satisfy all relations in R. If only one of these relations belongs to R^T, then variable x will assume a fixed value in all sets of signal values satisfying all the relations in R. For example, if R^T contains the relation $x \Rightarrow \overline{x}$ but does not contain the relation $\overline{x} \Rightarrow x$, then variable x can only assume the value 0. Since all relations in R have already been included in the energy function, only those relations in R^T that do not belong to R are added to the energy function. Any relation $x \Rightarrow y$ results in the addition of a term $A \times x \times (1 - y)$, where $A > 0$ is a constant, to the energy function. The computation of transitive closure can easily be accelerated through parallel processing [8].

8.4.2 Path Sensitization

A test must sensitize *at least* one path from the fault site to a primary output. Therefore, if a signal with no fanout is to be on a sensitized path, so must be the output of the gate driven by that signal. Similarly, if a fanout point is to be on a sensitized path, then at least one of the gates driven by the fanout must be on the sensitized path. This observation is based purely on circuit connectivity and is independent of the types of gates on the paths. Path sensitization information can be incorporated into the energy function [15].

Suppose that the AND gate in Figure 8.3 was embedded in a circuit. Assume that only signals a and c are on a path from the fault site to some primary output and signal a is not a fanout point. The case where a is a fanout point can be treated similarly. Let a' and c' be the corresponding faulty circuit signals. We introduce an additional binary variable s_i associated with signal i that lies on a potential path from the fault site to a primary output such that s_i assumes the value 1 whenever a path is sensitized through i. Clearly, $s_a = 1$ implies that a and a' assume opposite values but if $s_a = 0$, the two signals can assume arbitrary values. This relationship can be expressed by the following energy function:

$$F(V_a, V_{a'}, s_a) \; = \; s_a(V_a V_{a'} + (1 - V_a)(1 - V_{a'}))$$

Furthermore, for signal a to be on a sensitized path, signal c must also be on the sensitized path. This relationship can be expressed as follows:

$$S(s_a, s_c) \; = \; s_a(1 - s_c)$$

The expression $F(V_a, V_{a'}, s_a) + F(V_c, V_{c'}, s_c) + S(s_a, s_c)$ incorporates the path sensitization constraint for the single output gate.

The expression $S(s_a, s_c)$ can be written as a binary relation $s_a \Rightarrow s_c$ and, hence, is included in the set R. By adding such relations to R and using transitive closure, all *dominators* of the fault will be determined. Dominators are signals through which the fault effect must pass in order to reach an output [13].

8.5 ANZA Neurocomputer Results

We verified the feasibility of the proposed test generation technique by generating tests and identifying redundancies for several combinational circuits [6]. Two of them, the Schneider's circuit [16] and the SN5483A [18] four-bit carry look-ahead binary full adder, are included here. Tests for Schneider's circuit were generated using the ANZA neurocomputer but due to the limited memory on the ANZA co-processor board, tests for the four-bit adder could not be generated. Since the ANZA essentially solves a first order system of difference equations using the Euler method, we wrote a computer program to do the same. The program was implemented in the C language to run on a SUN 3/50 UNIX workstation.

A test for a fault using the ANZA co-processor board is generated as follows:

1. Construct the energy function for the circuit with the fault and incorporate additional constraints into the energy function using techniques described in Section 8.4. This is done on the host PC external to the ANZA co-processor board.

2. Load the neural network corresponding to the energy function onto the ANZA co-processor board using the system subroutines. Loading the network involves the transfer of link weights and thresholds of all neurons to the ANZA co-processor board.

3. Transfer the initial values for all neuron input voltages (i.e., u_i's) and the values of the step-size h and gain λ to the ANZA co-processor board using the system subroutines.

4. Supply the number of iterations (in the Euler method) to be performed by the ANZA co-processor.

5. Iterate the neural network until the pre-specified iteration limit is reached.

6. Convert the V_i's to the nearest binary digit, i.e., if $V_i \geq 0.5$, round off V_i to 1, and if $V_i < 0.5$, round off V_i to 0.

7. Check if the circuit energy function evaluates to 0 at these values of the V_i's. If the energy function is 0, a test has been obtained. If it is not 0, then the fault is either redundant or the neural network failed to find a test for the fault.

For all examples discussed here, the following values were used for the constants associated with the gates [7]: $A = 50$, $B = 50$ and $J = 50$. The gain λ was set to 50 and the u_i's (i.e., neuron inputs) were initialized to 0. This resulted in the V_i's being initialized to 0.5 (from Equation 8.3). Furthermore, the iteration limit was set to 1000. However, if none of the u_i's changed by more than 0.0001 in two successive iterations, the iteration process was terminated earlier.

Schneider's circuit [16]: The circuit used here is a slight modification of the original circuit that was used in Chapter 7. Of the three primary outputs, we retained only the one in the center of the circuit. This introduced four redundant faults.

We chose a fault and constructed the energy function. Only additional constraints generated using the transitive closure technique (Section 8.4) were included in the energy function. The transitive closure was determined using Warshall's algorithm [19]. The difference equations associated with the neural network corresponding to the energy function were solved using the ANZA co-processor board, with step size $h = 0.0001$. Due to the modification stated above, there are four redundant faults, all of which were identified through the transitive closure technique. Tests were generated for every detectable fault in the circuit. Although all test vectors were verified by fault simulation, faults fortuitously detected by a test vector were not dropped. The results obtained by the SUN 3/50 program were identical to the results obtained using the ANZA.

The choice of initial values for the V_i's was critical. For initial values of V_i's chosen from 0 and 1, the ANZA was unable to generate tests for several faults. This suggests the likelihood that the gradient of the energy function (Equation 4.7) near the center of the hypercube (state space) might be pointing toward a global minimum. We feel that this phenomenon, which has also been observed by other researchers [12], needs further investigation.

4-bit Adder [18]: Random vectors were generated for fault simulation until three successive vectors (an arbitrary choice) did not detect any additional fault. The random vector phase was followed by deterministic test generation based on the energy minimization technique. In this phase, we chose a fault from the remaining faults and constructed the corresponding energy function. The system of difference equations describing the neural network was solved using the SUN 3/50 program, with step size $h = 0.00001$. The values of the V_i's were rounded off to the nearest binary digit 0 or 1, and if Equation 4.2 evaluates to 0 at these values, we have a test for the fault. We performed fault simulation and we dropped fortuitously detected faults from further consideration.

Tests were generated for all but six faults. For each of the six undetected faults, the neural network settled in a local minimum of the energy function. The solution obtained by solving the system of difference equations is sensitive to the gain and the step size. Further investigation of the effect of varying the gain λ and the step size h should lead to useful results. The first order system described by Equation 4.6 is a stiff system since the exact solution contains the term $e^{-t/\tau}$, which decays to 0 with increasing t. Significant difficulties can occur when standard numerical techniques are applied to approximate the solution of such systems. Unless h has small values, the round-off error associated with the decaying transient portion can dominate the calculations and produce meaningless results.

As in Schneider's circuit, the choice of initial values for the V_i's was critical here as well. If the initial values of the V_i's were taken as 0 or 1, fewer faults were detected again suggesting that the gradient of the energy function at the center of the state-space might be oriented toward a global minimum.

The experimental results using the ANZA neurocomputer and the SUN 3/50 program are summarized in Tables 8.1 and 8.2, respectively. *RTG Vectors* is the number of test vectors obtained from random pattern test generation. *RTG Detected* is the number of faults detected by RTG vectors. Similarly, *DTG Vectors* is the number of test vectors generated by the energy minimization technique and *DTG Detected* is the number of faults thus detected.

TABLE 8.1: Experimental results using the ANZA neurocomputer.

Circuit	Schneider
Number of Inputs	4
Number of Outputs	1
Number of Gates	12
Total Faults	30
DTG Detected	26
DTG Vectors	6
Redundant Faults	4
Aborted Faults	0

TABLE 8.2: Experimental results using the SUN 3/50 program.

Circuit	Schneider	SN5483A
Number of Inputs	4	9
Number of Outputs	1	5
Number of Gates	12	60
Total Faults	30	152
RTG Detected	-	129
RTG Vectors	-	9
DTG Detected	26	23
DTG Vectors	6	10
Redundant Faults	4	-
Aborted Faults	0	6

8.6 Summary

The novel aspect of the approach discussed here is the application of a continuous optimization method for test generation, quite unlike the conventional discrete methods. Tests were generated by simulating analog neural networks on a commercial neurocomputer. Preliminary results on several combinational circuits confirm the feasibility of this technique. Due to the memory constraints on the neurocomputer, only small circuits were processed.

References

[1] V. D. Agrawal and S. C. Seth. *Test Generation for VLSI Chips*. IEEE Computer Society Press, Los Alamitos, CA, 1988.

[2] A. V. Aho, J. E. Hopcroft, and J. D. Ullman. *The Design and Analysis of Computer Algorithms*. Addison-Wesley Publishing Company, Reading, MA, 1974.

[3] J. Alspector and R. B. Allen. A Neuromorphic VLSI Learning System. In P. Loseleben, editor, *Advanced Research in VLSI: Proceedings of the 1987 Stanford Conference*, pages 313–349. MIT Press, Cambridge, Massachussetts, 1987.

[4] ANZA – Neurocomputing Coprocessor System. *Computer*, 21(9):99, September 1988.

[5] L. E. Atlas and Y. Suzuki. Digital Systems for Artificial Neural Networks. *IEEE Circuits and Devices Magazine*, 5(6):20–24, November 1989.

[6] S. T. Chakradhar, V. D. Agrawal, and M. L. Bushnell. On Test Generation Using Neural Computers. *International Journal of Computer Aided VLSI Design*, 3, March 1991.

[7] S. T. Chakradhar, M. L. Bushnell, and V. D. Agrawal. Automatic Test Pattern Generation Using Neural Networks. In *IEEE Proceedings of the International Conference on Computer-Aided Design*, pages 416–419, November 1988.

[8] D. Coppersmith and S. Winograd. Matrix Multiplication via Arithmetic Progressions. In *Proceedings of the 19th Annual ACM Symposium on Theory of Computing*, pages 1–6, May 1987.

[9] P. Hansen, B. Jaumard, and M. Minoux. A Linear Expected-Time Algorithm for Deriving All Logical Conclusions Implied by a Set of Boolean Inequalities. *Mathematical Programming*, 34(2):223–231, March 1986.

[10] J. J. Hopfield and D. Tank. Neural Computation of Decisions in Optimization Problems. *Biological Cybernetics*, 52(3):141–152, July 1985.

[11] M. L. James, G. M. Smith, and J. C. Wolford. *Applied Numerical Methods for Digital Computation*. Harper and Row Publishers, New York, NY, 1985.

[12] J. L. Johnson. A Neural Network Approach to the 3-Satisfiability Problem. *Journal of Parallel and Distributed Computing*, 6(2):435–439, February 1989.

[13] T. Kirkland and M. R. Mercer. A Topological Search Algorithm For ATPG. In *Proceedings of the 24th ACM/IEEE Design Automation Conference*, pages 502–508, June 1987.

[14] C. D. Kornfeld, R. C. Frye, C. C. Wong, and E. A. Rietman. An Optically Programmed Neural Network. In *Proceedings of the International Conference on Neural Networks, San Diego, CA*, volume 2, pages 357–364, July 1988.

[15] T. Larrabee. Efficient Generation of Test Patterns Using Boolean Difference. In *Proceedings of the IEEE International Test Conference*, pages 795–801, August 1989.

[16] P. R. Schneider. On the Necessity to Examine D-Chains in Diagnostic Test Generation. *IBM Journal of Research and Development*, 11(1):114, January 1967.

[17] B. Soucek and M. Soucek. *Neural and Massively Parallel Computers – The Sixth Generation*. John Wiley & Sons, Wiley Interscience publications, New York, NY, 1988.

[18] *The TTL Data Book for Design Engineers, Second edition*, page 199. Texas Instruments, 1973.

[19] S. Warshall. A Theorem on Boolean Matrices. *Journal of the ACM*, 9(1):11–12, January 1962.

Chapter 9

QUADRATIC 0-1 PROGRAMMING

"The first American high-speed computers, built in the early 1940s, used decimal arithmetic. But in 1946, an important memorandum by A. W. Burks, H. H. Goldstine, and J. von Neumann, in connection with the design of the first stored-program computers, gave detailed reasons for the decision to make a radical departure from tradition and to use base-two notation [see J. von Neumann, Collected Works 5, 41-65]. Since then binary computers have multiplied. After a dozen years of experience with binary machines, a discussion of the relative advantages and disadvantages of radix-2 notation was given by W. Buchholz in his paper 'Fingers or Fists?' [CACM 2, December 1959, 3-11]."

– D. E. Knuth in *The Art of Computer Programming*, Volume 2, Second Edition, Addison-Wesley (1981)

Once the test generation problem has been formulated as an optimization problem on a neural network, several methods can be used to find the minimum of the energy function. These can be classified as follows:

1. Use a single-processor or multiprocessor computer, or a hardware accelerator, to simulate the neural network.

2. Use an actual neural network [3, 17].

3. Use graph-theoretic or mathematical programming optimization techniques [10].

The first two methods were investigated in Chapters 7 and 8. In this chapter, we investigate graph-theoretic techniques for test generation. We present a new, discrete non-linear programming technique for test generation. The energy function for this problem has a special structure (Section 9.3) which we exploit to find an exact solution to the minimization problem. Knowledge specific to the solution is used to guide and accelerate energy minimization. This approach differs radically from classical algorithms [1].

Although several techniques have been proposed for minimizing quadratic functions (the energy function is quadratic) [9, 13], currently there is little hope to find exact solutions to even modest-sized problems [1] using these methods. The number of variables in the energy function arising from test generation problems is linear in the number of signals in the digital circuit. Therefore, it appears unlikely that traditional techniques can find exact solutions to energy minimization problems arising from circuits with more than about a hundred signals. New techniques that take advantage of the special structure of functions arising from a specific problem are needed. In this chapter, we develop a novel discrete non-linear programming technique for finding exact minimizers of energy functions arising from the test generation problem.

9.1 Energy Minimization

A formulation of the test generation problem as an energy minimization problem is discussed in Chapter 6 where the digital circuit was modeled as a neural network. For any given fault, we create a copy of the sub-circuit affected by the fault. The corresponding outputs of the *fault-free* circuit and the *faulty* circuit are passed through exclusive-OR gates. The outputs of these exclusive-OR gates are fed to an OR gate whose output is constrained to be 1. In addition, for a s-a-0(1) fault, the signals at the fault site in the fault-free and faulty parts are constrained to be 1(0) and 0(1), respectively. Any set of consistent signal values now corresponds to a test.

This formulation captures three *necessary* and *sufficient* conditions that any set of signal values must satisfy to be a test. First, the set of values must be consistent with each gate function in the circuit. Second, the two signals in the fault-free and faulty circuits at the fault site must assume

---

[1]Functions with less than 150 variables.

opposite values (e.g., 0 and 1, respectively, for a s-a-1 fault). Third, for the same primary input vector, the fault-free and faulty circuits should produce different output values.

In theory, the energy function resulting from the basic formulation (described in Chapter 6) can be minimized and a test generated. However, to make the minimization process efficient, we may use any prior knowledge about the solution. We first discuss the energy minimization procedure and then discuss enhancements to the basic formulation that accelerate the minimization process. We show that additional knowledge about the solution effectively restricts the search space for the energy minimum.

9.2 Notation and Terminology

Let $V = \{x_1, ..., x_n\}$ denote a set of Boolean (0-1) variables. The complement of the variable x_i is defined by

$$\overline{x_i} = 1 - x_i \qquad (9.1)$$

The elements of the set $L = \{x_1, \overline{x_1}, ..., x_n, \overline{x_n}\}$ are called *literals*. A *monomial* is a product aT where T is a term (i.e., a finite product of distinct literals) and a is a real number called the *coefficient* of the monomial.

A *pseudo-Boolean quadratic function* (PBQF) [13] $f : \{0, 1\}^n \Rightarrow R$ is defined as $f(\mathbf{x}) = \mathbf{x}^T \mathbf{Q} \mathbf{x} + \mathbf{c}^T \mathbf{x}$, where $\mathbf{Q}$ is a symmetric $n \times n$ matrix with null elements in the diagonal, $\mathbf{c}$ is a vector of n real numbers, $\mathbf{x}$ is a vector of n binary 0-1 variables and $\mathbf{x}^T$ is the transpose of $\mathbf{x}$. There is no loss of generality due to the null diagonal assumption because $x_i^2 = x_i$ for $1 \leq i \leq n$. The *quadratic 0-1 minimization* problem is to find the minimum in $\{0, 1\}^n$ of the pseudo-Boolean quadratic function.

A *positive pseudo-Boolean form* (*posiform*) is any sum of monomials with positive coefficients. Any PBQF can be written as a posiform. This is a direct consequence of the identity

$$-x_i x_j = \overline{x_i} x_j + \overline{x_j} - 1 \qquad (9.2)$$

However, a PBQF may not have a unique posiform representation. For example, the PBQF $f = 3 + x_1 - 2x_1 x_2$ can be written in posiform as $f = 2 + \overline{x_1} + 2x_1 \overline{x_2}$ or $f = 1 + x_1 + 2\overline{x_2} + 2\overline{x_1} x_2$. A posiform without a constant term is *homogeneous* and a posiform with a constant term is *inhomogeneous*.

The problem of minimizing or maximizing a pseudo-Boolean quadratic function is NP-complete [13]. However, several polynomial time solvable special classes of PBQF's have been identified. The following result is essential to the present discussion: *A minimizing point of a quadratic ho-mogeneous posiform, known to have a minimum value 0, can be obtained in time complexity that is linear in the number of variables and terms in the function* [6]. We exploit this result to efficiently minimize the energy function.

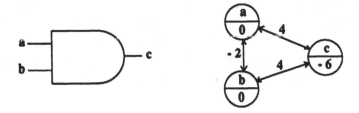

FIGURE 9.1: An AND gate (a) and its neural net model (b).

9.3 Minimization Technique

Unlike the general quadratic 0-1 programming problems, the quadratic function arising from a test generation problem has a special structure. For example, consider the AND gate shown in Figure 9.1. Its energy function is derived from Table 5.1 by setting $A = B = 2$ to give[2]:

$$E_{AND}(a, b, c) = [2c(1 - a)] + [2c(1 - b)]$$
$$+[2c - 2ac - 2bc + 2ab] \qquad (9.3)$$

All three terms are non-negative and, hence, $E_{AND} = 0$ requires that each term should simultaneously become 0. Rewriting the first two terms in posiform, using Equation 9.1, we get:

$$E_{AND}(a, b, c) = [2c\bar{a}] + [2c\bar{b}] + [2c - 2ac - 2bc + 2ab]$$

Observe that the sub-function comprising the first two terms is a homoge-neous posiform with a minimum value of 0. The third term can be rewritten only as an inhomogeneous posiform using Equation 9.1 and Equation 9.2: $2\bar{a}c + 2\bar{b}c + 2ab + 2\bar{c} - 2$. Although monomials $2\bar{a}c$, $2\bar{b}c$, $2ab$ and $2\bar{c}$ have positive coefficients, we do not include them in the homogeneous posiform. This is because the inclusion of these monomials results in a homogeneous posiform with a minimum value other than 0 (in our exam-ple, the homogeneous posiform will assume a minimum value of 2) and

---

[2]For simplicity in writing expressions, we use signal names to also denote activation values of corresponding neurons. Thus, the value V_a of neuron a is simply denoted as a.

the quadratic homogeneous posiform minimization technique we alluded to in the previous section is no longer applicable.

The energy functions for the other logic gate types (Table 5.1) have a similar structure. Therefore, E_{CKT}, the energy function for the entire circuit, can be expressed as two sub-functions: a homogeneous posiform E_H and an inhomogeneous posiform E_I, each with a minimum value of 0. Note that all circuit signals appear in the homogeneous posiform. This immediately suggests the following method to minimize the energy function:

1. Find a new minimizing point α of E_H. This point can be obtained in time complexity that is linear in the number of variables and terms in E_H [6].

2. Check if α is also a minimizing point of E_I. If it is, then α is a minimizing point of E_{CKT} and we have found a test. Otherwise go to Step 1.

This approach drastically reduces the required computing because we solve a greatly simplified homogeneous posiform and check the solution against the inhomogeneous posiform. Larrabee recently used a similar approach but formulated the test generation problem as a Boolean satisfiability problem [18]. The homogeneous posiform may have exponentially many minimizing points. We use a branch-and-bound technique to systematically cycle through these points to find a minimizing point of E_{CKT}.

To explain our method, we require an understanding of the *implication graph* that represents signal dependencies and the topology of the circuit. In Figure 9.2, observe that if b is 1 then $\bar{a}$ must be true and if c is 1 then both

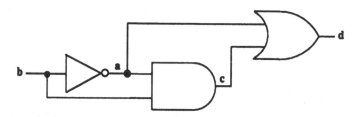

FIGURE 9.2: An example circuit.

a and b must be 1. The graph in Figure 9.3 represents these implications as arcs from b to $\bar{a}$, from c to a and from c to b. Normally, there would be additional arcs, e.g., a to d and c to d, but these are deleted from the graph because we assume that signal d is assigned a fixed value and, hence, it no longer depends on a and c.

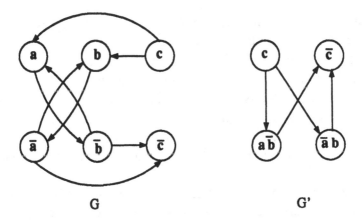

FIGURE 9.3: Implication graph G and its condensation G'.

The implication graph can also be derived directly from the posiform. A minimizing point of the homogeneous posiform can be obtained as follows [4]:

1. Construct a directed implication graph [4] $G(L, \mathcal{E})$, where L is the vertex set and $\mathcal{E}$ is the arc set. The arc set $\mathcal{E}$ is such that associated with each monomial $z x_i x_j$ (z is the coefficient) in the posiform are two arcs $(x_i, \overline{x_j})$ and $(x_j, \overline{x_i})$. An arc (x_i, x_j) expresses the relationship $x_i \Rightarrow x_j$ (i.e., if literal $x_i = 1$ then $x_j = 1$, but if $x_i = 0$, x_j could be either 0 or 1). As an example, consider the homogeneous posiform E_H given by Equation 9.5. Monomial $2ab$ contributes arcs $(a, \overline{b})$ and $(b, \overline{a})$ to the implication graph of E_H which is shown in Figure 9.3. Arcs corresponding to the other monomials in E_H can be derived similarly. The graph G has a *duality* property [4]: it is isomorphic to the graph obtained from G by reversing the directions of all arcs and complementing all literals.

2. Find all strongly-connected components of G [2] and construct the *condensation* graph $G'(L', \mathcal{E}')$ as follows. There is a vertex in L' for each strongly-connected component of G and let vertices c_i and c_j correspond to the ith and jth strongly-connected components of G. There is an arc from c_i to c_j if there is an arc in $\mathcal{E}$ from some vertex in the ith component to a vertex in the jth component. By duality, every strong component i has a dual strong component $dual(i)$ consisting of the subgraph induced by the complements of the literals in component i. Similarly, define $c_i = dual(c_j)$ if component i is the dual of component j. Observe that if a literal in a strongly-connected component is set to 1 (0), so are all literals in the same (dual) strongly-

connected component. As an example, the implication graph of E_H (Figure 9.3) has four strongly-connected components: $\{\overline{c}\}$, $\{\overline{a}, b\}$, $\{a, \overline{b}\}$ and $\{c\}$. The condensation of the implication graph is shown in Figure 9.3. Furthermore, if $a = 1$ then $\overline{b} = 1$, since arc $(a, \overline{b})$ expresses the relationship $a \Rightarrow \overline{b}$. If $a = 0$ then $\overline{a} = 1$ and due to the arc $(\overline{a}, b)$, $b = 1$.

3. Initially, every vertex in G' is *unmarked*. Process the vertices of G' in a topological order as follows: If c_i is marked, do nothing. Otherwise, if $i = dual(i)$, then stop since the homogeneous posiform does not have a minimum value of 0 [4]. Otherwise, mark c_i and $dual(c_i)$, set all literals in the ith component to 0 and assign all literals in $dual(i)$ the value 1. An example in the next section will illustrate this method.

Each of the above steps can be performed in time complexity that is linear in the number of vertices and arcs in G [4].

To systematically enumerate all minimizing points of E_H, for the worst case, one may have to examine all 2^n combinations of variables in E_H. However, because of the duality property of G, we need not examine all variable combinations.

Size of search space: The following remarks are a direct consequence of the duality property of graph G:

- Every strong component has a dual.

- If a literal x_l in strong component i is assigned the value 1(0), then all literals in the component i are set to 1(0) and all literals in $dual(i)$ are set to 0(1).

Therefore, for every strong component and its dual, we only have to assign a value to a single literal. If G has w strong components, we can enumerate all minimizing points of E_H by examining only $2^{\frac{w}{2}}$ combinations of variable values.

Variable expansion order: In general, the graph G' can have several topological orderings and if E_H has a minimum value of 0, each ordering may result in a different minimizing point. Unfortunately, by considering all topological orderings and generating a minimizing point for each by the above method, we may not cover all minimizing points of E_H. Consequently, we suggest the following heuristic method: Select one ordering from all possible topological orderings. This ordering automatically

imposes a total order on the decision variables. Use a branch-and-bound
procedure to systematically consider all combinations of decision variables
and generate all minimizing points of E_H.

9.4 An Example

Consider the problem of generating a test for the s-a-0 fault on signal d in
the circuit of Figure 9.2. Since the fault is on a primary output, we need
not create a faulty circuit copy. We constrain d to assume the value 1 to
sensitize the fault. From Table 5.1 (with $A = B = 2$):

$$
\begin{aligned}
E_{NOT}(b, a) &= [2ab] + [2\bar{a}\bar{b}] \\
E_{AND}(a, b, c) &= [2\bar{a}c] + [2\bar{b}c] + [2c - 2ac - 2bc + 2ab] \\
E_{OR}(a, c, d) &= [2a\bar{d}] + [2c\bar{d}] + [2d - 2cd - 2ad + 2ac]
\end{aligned}
$$

The energy function for the circuit is:

$$
E_{CKT} = E_{NOT}(b, a) + E_{AND}(a, b, c) + E_{OR}(a, c, d)
$$

and the homogeneous posiform E_H is:

$$
E_H = 2ab + 2\bar{a}\bar{b} + 2\bar{a}c + 2\bar{b}c + 2a\bar{d} + 2c\bar{d} \qquad (9.4)
$$

The remaining terms in E_{CKT} can be rewritten as an inhomogeneous posiform using Equation 9.1 and Equation 9.2. However, there is no particular
advantage in doing this. Our method does not require that these terms be
explicitly represented as an inhomogeneous posiform since we only have
to verify whether these terms evaluate to 0 at a minimizing point of E_H.
Substituting $d = 1$, E_H can be simplified to:

$$
E_H = 2ab + 2\bar{a}\bar{b} + 2\bar{a}c + 2\bar{b}c \qquad (9.5)
$$

The corresponding implication graph G and the condensation of G are
shown in Figure 9.3. A reverse topological ordering of strong components
in G is as follows: $\{\bar{c}\}$, $\{\bar{a}, b\}$, $\{a, \bar{b}\}$ and $\{c\}$. The duals of components
$\{\bar{c}\}$ and $\{\bar{a}, b\}$ are $\{c\}$ and $\{a, \bar{b}\}$, respectively. Selecting one variable from
every strong component and its dual, a possible variable expansion order
for the branch-and-bound process is: c, a.

Let us first generate a minimizing point using the method of Aspvall
et al [4]. First, mark components $\{c\}$ and its dual $\{\bar{c}\}$, and set $c = 0$ and
$\bar{c} = 1$. Then, mark components $\{a, \bar{b}\}$ and its dual $\{\bar{a}, b\}$, and set $a = 0$,
$\bar{b} = 0, \bar{a} = 1$ and $b = 1$. Components $\{a, \bar{b}\}$ and $\{c\}$ are already marked.

Therefore, a minimizing point of E_H is $a = c = 0$ and $b = 1$. At these values of a, b and c, $E_H = 0$ but E_I does not evaluate to 0 and we must look for another minimizing point of E_H.

All minimizing points of E_H can be systematically generated by a branch-and-bound search on the variables c and a. If $c = 1$, then literals in components $\{a, \bar{b}\}$ and $\{\bar{a}, b\}$ should be set to 1. However, this causes a conflict because a variable and its complement are being assigned the same value. Thus, there is no minimizing point of E_H with $c = 1$. We now set $c = 0$ which does not imply any other variable. Assign $a = 0$ and set all literals in the component containing $\bar{a}$ to 1. Since each literal in a component implies all other literals in that component, all literals in the same component must be 1 when any one literal is set to 1. This results in $b = 1$ and the minimizing point is $a = c = 0$, $b = 1$. Unfortunately, this also is not a minimizing point of E_I. We backtrack and re-assign $a = 1$. This results in $b = 0$ and the minimizing point $a = 1$, $b = c = 0$ which is a minimizing point of E_I. Therefore, $b = 0$ is a test vector for the fault d s-a-0.

From the example, we observe the following:

1. In all minimizing points of E_H, $c = 0$. In general, one or more variables may assume fixed values at all minimizing points. Identifying such variables and their respective values would greatly reduce the search space. Fixing each variable halves the search space. Section 9.5.1 discusses a technique to identify all fixed variables.

2. Had we tried the choice $a = 1$ before $a = 0$, we would have obtained a test earlier. Given a set of values for the decision variables, a simple heuristic to decide which value to try first for the next decision variable would be as follows: Let $n_1(n_0)$ be the number of terms in E_I that evaluate to 0 when a certain variable a assumes the value $1(0)$. If $n_1 > n_0$ then try the choice $a = 1$ first.

3. Since $d = 1$, the term $[2d - 2cd - 2ad + 2ac]$ in E_I becomes $[2 - 2a - 2c + 2ac]$ which is a homogeneous posiform $[2\bar{a}\,\bar{c}]$ (apply Eqns. 9.1 and 9.2 to see this). Include this term in the implication graph G for E_H and we have only two strong components: $\{a, \bar{b}, \bar{c}\}$ and its dual $\{\bar{a}, b, c\}$. Selecting one variable for every component and its dual, the branch-and-bound process now only has c as the decision variable. In general, fixing the values of variables may make some terms in E_I homogeneous and we can include these terms in E_H. Section 9.5.2 discusses a technique to move terms from E_I to E_H. This is advantageous because it eliminates decision variables and

because the minimization of E_H can be performed in linear time.

4. All topological orderings of the strong components of G give the same variable expansion order (c and a are the first and second decision variables, respectively). In general, different topological orderings may lead to several different variable expansion orders, any one of which can be selected.

9.5 Accelerated Energy Minimization

The minimization process can be speeded up in several ways:

1. Reduce the number of decision variables in the branch-and-bound process. Section 9.5.1 discusses a technique using transitive closure to fix the largest number of variables in E_{CKT}.

2. Reduce the number of minimizing points of the homogeneous posiform E_H. Prior knowledge about the solution can be used to determine additional constraints on the variables in E_H which may decrease the number of minimizing points that must be considered. Section 9.5.2 discusses a technique to identify additional constraints.

3. Prune out the non-solution branches of the decision tree as much as possible. In Section 9.5.3 we use problem-specific knowledge to eliminate several minimizing points of the homogeneous posiform E_H that would not have minimized E_I.

4. Order the decision variables such that minimizing points with higher likelihood of being a test are examined earlier.

9.5.1 Transitive Closure

We present a technique to identify the variables in E_H that assume constant values at all minimizing points of E_H. The implication graph G essentially represents a set R of m binary relations, where m is the number of arcs in the graph. R may imply the following types of logical conclusions:

1. *Contradiction*: Some variable x_i must assume both 0 and 1 values. A contradiction occurs if a strongly-connected component of G contains both x_i and $\overline{x_i}$ for some $i \in \{1, 2, ..., n\}$ [4]. A contradiction implies that the homogeneous posiform does not have a minimum value of 0. *This occurs in the case of a redundant fault.*

FIGURE 9.4: An implication graph showing contradiction.

	a	$\bar{a}$	b	$\bar{b}$	c	$\bar{c}$
a	1	0	0	1	0	1
$\bar{a}$	0	1	1	0	0	1
b	0	1	1	0	0	1
$\bar{b}$	1	0	0	1	0	1
c	1	1	1	1	1	1
$\bar{c}$	0	0	0	0	0	1

FIGURE 9.5: Adjacency matrix of G_T.

2. *Identification*: Certain pairs of literals (x_i, x_k) must assume the same value. All literals in a strongly-connected component must assume the same value.

3. *Fixation*: Some variable x_i can be fixed to the value 0 or 1.

By finding strongly-connected components of G [22], one can detect any possible contradictions and perform all identifications in $O(m)$ time. All fixations implied by R can be determined by finding the *transitive closure* [2] of graph G. The graph G_T, the transitive closure of G, has the same vertex set as G, but has an arc from x_i to x_j if, and only if, there is a directed path (of length 0 or longer) from x_i to x_j in G. A contradiction occurs if G_T has both arcs $(x_i, \overline{x_i})$ and $(\overline{x_i}, x_i)$. Figure 9.4 shows the two arcs. Since x_i and $\overline{x_i}$ belong to the same strongly-connected component, they must be identically labeled, which is impossible. A fixation results if G_T has only one of the arcs. If arc $(x_i, \overline{x_i})$ ($(\overline{x_i}, x_i)$) is present in G_T, fix $x_i = 0$ ($x_i = 1$). As an example, consider the graph G in Figure 9.3. The strongly-connected components are $\{c\}, \{\bar{c}\}, \{a, \bar{b}\}$ and $\{\bar{a}, b\}$. Since none of the components contain a variable and its complement, there is no contradiction. Since literals a and $\bar{b}$ belong to the same strongly-connected component, they must assume the same value. Similarly, $\bar{a}$ and b must assume the same value. The *adjacency matrix* [2] of G_T, the transitive closure of G, is shown in Figure 9.5. The element in row x_i and column x_j is 1 if arc (x_i, x_j) is in G_T and 0 otherwise. We can fix $c = 0$ since arc $(c, \bar{c})$ is in G_T.

Finding G_T requires $O(n^3)$ operations in the worst case [24] (n is the number of vertices in G). This complexity could be reduced to $O(n^{2.49})$

by exploiting the equivalence between matrix multiplication and transitive closure and using fast matrix multiplication methods. An alternate method for deriving all logical conclusions in linear expected-time complexity is given by Hansen *et al* [14]. This method avoids computing the complete transitive closure but may not be easily parallelizable.

The computation of transitive closure can easily be accelerated through parallel processing. Transitive closure belongs to class *NC*, a hierarchy of problems solvable by deterministic algorithms that operate in polylog time using a polynomial-bounded number of processors. If A is the adjacency-matrix of graph G, the transitive closure of G can be obtained by squaring the adjacency matrix $\log_2 n$ times, where n is the number of vertices in graph G. It is known [11] that $n \times n$ matrix multiplication can be performed in $O(\log_2 n)$ time using $O(n^{2.376})$ processors. Thus, transitive closure can be computed in polylog time using the same number of processors. If fewer processors are available, we can use *Brent's scheduling principle* [7] to design an efficient parallel algorithm. From a practical perspective, recent implementations given by Golub and Van Loan [12] and Bailey [5] are significant.

9.5.2   Additional Pairwise Relationships

In general, given a set of fixed values for some variables, certain terms in E_I become homogeneous and hence can be included in E_H. When a new term is added to E_H, the transitive closure of G can be incrementally updated to find any additional fixations.

In the example of Section 9.4, we noticed that since $d = 1$, one term in the inhomogeneous posiform changes to the homogeneous posiform $2\overline{a}\,\overline{c}$. From the transitive closure of G, c was fixed at 0. Since $2\overline{a}\,\overline{c}$ must assume the value 0, we must fix $\overline{a} = 0$. This fixes all literals in the strongly-connected component containing $\overline{a}$ to 0 (e.g., b becomes 0) and the values of all variables are determined without any search. The minimizing point is $a = 1$, $b = c = 0$. E_I evaluates to 0 and the test vector for d s-a-0 is $b = 0$.

More constraints on variables in E_H can also be determined by repeating the following procedure for each unassigned literal x_i in the implication graph of E_H:

1. Fix x_i to 1.

2. Set all literals implied by x_i to 1. Such literals can easily be determined either from the transitive closure of the implication graph or by finding all descendants of x_i in the implication graph.

3. Check whether any term in E_I changes to homogeneous posiform. If so, include the appropriate arcs in the implication graph, update the transitive closure and determine new fixations, if any.

Again, set $d = 1$ in the example of Section 9.4. By constructing the implication graph G for the unsimplified E_H, as given by Equation 9.4, it is easily seen that there are no other literals in the strongly-connected component containing d and hence, no literal can be assigned a value. However, one term (as shown in Section 9.4) in E_I changes to homogeneous posiform $2\overline{a}\,\overline{c}$. This term results in the addition of arcs $(\overline{a}, c)$ and $(\overline{c}, a)$ to the implication graph. We appropriately update rows $\overline{a}$ and $\overline{c}$ in the transitive closure of G.

Such arcs added to the implication graph put more constraints on the variables in E_H and may reduce the number of minimizing points of E_H. In turn, the additional constraints may result in new fixations and reduce the search space. A technique to determine such constraints using circuit structure is given by Schulz et al [21].

9.5.3 Path Sensitization

The ideas of this section were previously discussed in Chapter 8 to augment the energy function. A test sensitizes *at least* one path from the fault site to a primary output. Therefore, if a signal with no fanout is to be on a sensitized path, so must be the output of the gate driven by this signal. Similarly, if a fanout point is to be on a sensitized path, at least one of the gates driven by the fanout must also be on the sensitized path. This observation is based purely on circuit connectivity and is independent of the types of gates in the sensitized paths. Path sensitization information can be incorporated into the energy function [18].

Suppose that the AND gate of Figure 9.1 is embedded in a circuit. Assume that only signals a and c are on a path from the fault site to some primary output and signal a is not a fanout point. The case where a is a fanout point can be treated similarly. Let a' and c' be the corresponding faulty circuit signals. We introduce an additional binary variable s_i associated with signal i that lies on a potential path from the fault site to a primary output such that s_i assumes the value 1 whenever a path is sensitized through i. Clearly, $s_a = 1$ implies that a and a' assume opposite values but if $s_a = 0$, the two signals can assume arbitrary values. This relationship can be expressed by the following energy function:

$$F(a, a', s_a) = s_a(aa' + (1 - a)(1 - a'))$$

Furthermore, for signal a to be on a sensitized path, signal c must also be on the sensitized path. This relationship can be expressed as follows:

$$S(s_a, s_c) \; = \; s_a(1 - s_c)$$

The expression $F(a, a', s_a) + F(c, c', s_c) + S(s_a, s_c)$ incorporates the path sensitization constraint for the single output gate.

The expression $S(s_a, s_c)$ is a homogeneous posiform and hence, can be included in E_H. By adding such terms to E_H and by using transitive closure, all *dominators* of the fault can be determined [19]. Dominators are signals through which the fault effect must pass in order to reach an output [16].

Classical test generation algorithms backtrack when the current set of signal assignments is such that there is no potentially sensitizable path from the fault site to any primary output. By adding path sensitization terms to E_{CKT}, it can be shown that the branch-and-bound process will automatically backtrack when the current partial assignment of variables is such that there is no potentially sensitizable path from the fault site to any primary output. Essentially, we detect the violation of a global constraint by examining several local constraints. This is particularly useful since the local constraints can be quickly examined in parallel.

9.6 Experimental Results

We verified the feasibility and efficiency of the proposed technique by generating tests and identifying redundancies for several combinational circuits. Three of them, the Schneider's circuit [20], the SN5483A [23] four-bit carry look-ahead binary full adder and the SN54LS181 [23] four-bit Arithmetic Logic Unit (ALU), are included here. The Schneider's circuit is modified, as in Chapter 8, to have a single output and this modification introduces four redundant faults.

A preliminary version of the test generation system was implemented in C to run on a SUN 3/50 workstation. Excepting the energy minimization part, this system is similar to that described in the previous chapters. The first phase of the system performs random test generation. Random vectors are generated for fault simulation until three successive vectors do not detect any additional fault. All vectors that did not improve fault coverage are dropped. The random phase is followed by deterministic test generation based on the energy minimization technique. In this phase, we choose a fault from the remaining faults and construct E_H and E_I. Enhancements mentioned in Section 9.5.1 and Section 9.5.2 were included but path sensitization constraints (Section 9.5.3) were not implemented in the present version. The transitive closure of E_H was determined using Warshall's algorithm [24] and fixations were then obtained. This was followed by a systematic enumeration of the minimizing points of E_H until a test vector was generated. Fault simulation was performed after every test vector to eliminate other detected faults.

The program generated tests for all testable faults in the three circuits and identified all redundant faults. Table 9.1 summarizes the results. *RTG Vectors* is the number of test vectors obtained from random test generation. *RTG Detected* is the number of faults detected by RTG vectors. Similarly, *DTG Vectors* is the number of test vectors generated by the energy minimization technique and *DTG Detected* is the number of faults thus detected. There are four redundant faults in the Schneider's circuit and four in the ALU, all of which were identified through contradiction in implication graphs of E_H. Hence, no search was conducted for identifying redundant faults. Furthermore, the branch-and-bound procedure generated tests for all detectable faults in the three circuits without any backtracks. *ATG Time* does not include the time for random pattern test generation and fault simulation. All vectors were verified via fault simulation.

We recommend that path sensitization constraints be incorporated into E_{CKT} to generate tests for larger circuits like the ISCAS combinational benchmark circuits [8].

TABLE 9.1: Experimental results.

Circuit	Schneider	SN5483A	SN54LS181
Number of Inputs	4	9	14
Number of Outputs	1	5	8
Number of Gates	12	60	132
Total Faults	30	152	263
RTG Detected	0	129	252
RTG Vectors	0	9	35
DTG Detected	26	23	7
DTG Vectors	6	10	6
Redundant Faults	4	0	4
Number of Backtracks	0	0	0
Total ATG Time (cpu sec.)	< 1	< 1	< 1

9.7 Summary

With the energy minimization formulation, a whole suite of discrete optimization techniques, different from the classical discrete techniques [1], can be used for test generation. We have demonstrated a new algorithm for generating test vectors based on quadratic 0-1 programming. Continuous optimization techniques using neural networks (Chapter 8) or those based on Karmarkar's interior point method for 0-1 integer programming [15] are other possibilities. We have shown how circuit-specific knowledge can be encoded in the energy formulation and graph-theoretic methods like transitive closure can be used to accelerate energy minimization. A detailed exposition of the use of transitive closure for test generation is given in the next chapter. Results for test generation and redundancy identification for several circuits confirm the feasibility of this technique. We believe that our method offers the potential for using parallel processing for test generation.

References

[1] V. D. Agrawal and S. C. Seth. *Test Generation for VLSI Chips*. IEEE Computer Society Press, Los Alamitos, CA, 1988.

[2] A. V. Aho, J. E. Hopcroft, and J. D. Ullman. *The Design and Analysis of Computer Algorithms*. Addison-Wesley Publishing Company, Reading, MA, 1974.

[3] J. Alspector and R. B. Allen. A Neuromorphic VLSI Learning System. In P. Loseleben, editor, *Advanced Research in VLSI: Proceedings of the 1987 Stanford Conference*, pages 313–349. MIT Press, Cambridge, Massachussetts, 1987.

[4] B. Aspvall, M. F. Plass, and R. E. Tarjan. A Linear-Time Algorithm for Testing the Truth of Certain Quantified Boolean Formulas. *Information Processing Letters*, 3(8):121–123, March 1979.

[5] D. Bailey. Extra High Speed Matrix Multiplication on the Cray-2. *SIAM Journal on Scientific and Statistical Computing*, 9(3):603–607, May 1988.

[6] A. Billionnet and B. Jaumard. A Decomposition Method for Minimizing Quadratic Pseudo Boolean Functions. *Operations Research Letters*, 8(3):161–163, June 1989.

[7] R. P. Brent. The Parallel Evaluation of General Arithmetic Expressions. *Journal of the ACM*, 21(2):201–206, April 1974.

[8] F. Brglez and H. Fujiwara. A Neutral Netlist of 10 Combinatorial Benchmark Circuits and a Target Translator in FORTRAN. In *Proceedings of*

the IEEE International Symposium on Circuits and Systems, pages 663–698, June 1985.

[9] M. W. Carter. The Indefinite Zero One Quadratic Problem. *Discrete Applied Mathematics*, 7(1):23–44, January 1984.

[10] S. T. Chakradhar, V. D. Agrawal, and M. L. Bushnell. Automatic Test Generation Using Quadratic 0-1 Programming. In *Proceedings of the 27th ACM/IEEE Design Automation Conference*, pages 654–659, June 1990.

[11] D. Coppersmith and S. Winograd. Matrix Multiplication via Arithmetic Progressions. In *Proceedings of the 19th Annual ACM Symposium on Theory of Computing*, pages 1–6, May 1987.

[12] G. H. Golub and C. F. Van Loan. *Matrix Computations*. Johns Hopkins Series in Mathematical Science, The Johns Hopkins University Press, Baltimore, Maryland, 1989.

[13] P. L. Hammer and B. Simeone. Quadratic Functions of Binary Variables. Technical Report RRR # 20-87, Rutgers Center for Operations Research (RUTCOR), Rutgers University, NJ 08903, June 1987.

[14] P. Hansen, B. Jaumard, and M. Minoux. A Linear Expected-Time Algorithm for Deriving All Logical Conclusions Implied by a Set of Boolean Inequalities. *Mathematical Programming*, 34(2):223–231, March 1986.

[15] N. Karmarkar, M. G. Resende, and K. G. Ramakrishnan. An Interior Point Approach to NP-Complete Problems. *Contemporary Mathematics*, 114:297–308, 1990.

[16] T. Kirkland and M. R. Mercer. A Topological Search Algorithm For ATPG. In *Proceedings of the 24th ACM/IEEE Design Automation Conference*, pages 502–508, June 1987.

[17] C. D. Kornfeld, R. C. Frye, C. C. Wong, and E. A. Rietman. An Optically Programmed Neural Network. In *Proceedings of the International Conference on Neural Networks, San Diego, CA*, volume 2, pages 357–364, July 1988.

[18] T. Larrabee. Efficient Generation of Test Patterns Using Boolean Difference. In *Proceedings of the IEEE International Test Conference*, pages 795–801, August 1989.

[19] S. R. Pawagi, P. S. Gopalakrishnan, and I. V. Ramakrishnan. Computing Dominators in Parallel. *Information Processing Letters*, 24(4):217–221, 1987.

[20] P. R. Schneider. On the Necessity to Examine D-Chains in Diagnostic Test Generation. *IBM Journal of Research and Development*, 11(1):114, January 1967.

[21] M. H. Schulz, E. Trischler, and T. M. Sarfert. SOCRATES: A Highly Efficient Automatic Test Pattern Generation System. *IEEE Transactions on Computer-Aided Design*, 7(1):126–136, January 1988.

[22] R. E. Tarjan. Depth-First Search and Linear Graph Algorithms. *SIAM Journal of Computing*, 1(2):146–160, June 1972.

[23] *The TTL Data Book for Design Engineers, Second edition*, page 199. Texas Instruments, 1973.

[24] S. Warshall. A Theorem on Boolean Matrices. *Journal of the ACM*, 9(1):11–12, January 1962.

Chapter 10

TRANSITIVE CLOSURE AND TESTING

"Once you show that a problem is equivalent to the 'chicken and egg problem' you do not have to solve it."

In previous chapters, we used transitive closure to speed up the energy minimization algorithms. Now we present a test generation algorithm entirely based on transitive closure. A test is obtained by determining signal values that satisfy a Boolean expression constructed from the circuit netlist and the fault. The algorithm is a sequence of two main steps that are repeatedly executed: transitive closure computation and decision-making. The transitive closure computation determines all logical consequences of any partial set of signal assignments. To compute the transitive closure of the circuit, we construct an implication graph whose vertices are labeled as the true and false states of all signals. A directed edge (x, y) in this graph represents the controlling influence of the true state of signal x on the true state of signal y. The signals x and y are connected through a wire or a gate in the circuit. Since the implication graph only includes pairwise (or binary) relations, it is a partial representation of the netlist. The transitive closure of the implication graph contains pairwise logical relationships among all signals. When signal relationships describing fault activation and path sensitization are included, transitive closure determines signal fixations and logical contradictions that directly identify many redundancies. Implication, unique path sensitization, static and dynamic learning, sensitization of physical and logical dominators and other techniques that are useful in determining necessary signal assignments are implicit in the process. If signals thus determined satisfy the Boolean formula, we have a test. Otherwise, we use the decision-making step, fix an unassigned signal, and update the transitive closure to determine all logical consequences of this decision. Since computation of transitive closure is as easily parallelizable as matrix multiplication, our algorithm is suitable for execution on a multiprocessor system.

10.1 Background

Some of the best known algorithms for test generation published to date use various types of analyses to speed up the basic branch-and-bound search process. For example, in the FAN algorithm, Fujiwara and Shimono [8] use multiple backtrace to quickly discover any possible conflicts. In the TOPS algorithm [12], Kirkland and Mercer rely on prior identification of *dominators*. In SOCRATES [17, 18], Schulz *et al* employ static and dynamic learning largely based on local simulation. More recently, in the EST algorithm [9], Giraldi and Bushnell store groups of signal states for future use. Rajski and Cox use reduction lists to quickly determine necessary assignments [16].

These algorithms solve two main problems: (1) they determine the order in which decisions on fixing signals should be made and (2) they determine logical consequences of a partial set of signal assignments. Although they make extensive use of the circuit structure and function, they have several shortcomings. First, they do not guarantee the identification of *all* logical consequences of a partial set of signal assignments. In other words, local conditions are easier to identify than the global ones. It is, however, desirable to identify all such consequences so that a branch-and-bound method can effectively avoid signal assignments that will not lead to a test vector. This may also result in an earlier detection of redundant faults. Second, they do not establish the complexity of determining the logical consequences and it is unclear if it is at all possible to determine all the consequences using reasonable amount of resources. Third, these techniques may not be easily parallelizable.

Parallelization of PODEM [10] was considered by Patil and Banerjee [14]. With their method, since the entire circuit resides in every processor of the multiprocessor system, it may be difficult to generate tests for circuits that do not fit on a single processor due to limited memory resources. Recently proposed energy minimization (Chapter 9) and Boolean satisfiability methods [13] seem more suitable for parallelization and some attempts to implement these methods on multiprocessor systems have been reported [19]. However, in all these methods, identification of logical consequences of a partial set of signal assignments is still an essential subproblem and a successful implementation will require an efficient parallel solution to this subproblem.

Although we suggest the use of transitive closure in earlier chapters, a complete test generation algorithm entirely based on transitive closure computations has never been described. In this chapter, we explore the application of transitive closure to solve the test generation problem [4]. We represent the digital circuit as a set of binary and ternary relations (*i.e.*, relations involving two and three binary variables), constructed from the circuit netlist and the fault specification. Any set of signal values that satisfies these relations will correspond to a test. The algorithm is a sequence of two main steps that are repeatedly executed: transitive closure computation and decision-making. The transitive closure computation identifies *all* logical consequences of a partial assignment of signal values.

The transitive closure of the digital circuit is computed from the implication graph whose vertices are the true and complemented states of the signals. A directed edge $(x, \overline{y})$ in this graph represents the controlling influence of the true state signal x on the false state of signal y. The variables x and y are associated with the same gate or signal net in the circuit. Since the implication graph includes only pairwise (or binary) relations, it is a partial representation of the netlist. The transitive closure of the implication graph contains pairwise logical relationships among all signal-pairs. When signal relationships describing fault activation and path sensitization are included, transitive closure determines signal fixations and logical contradictions that directly identify many redundancies. If signals thus determined satisfy the Boolean formula, we have a test. Otherwise, we enter the decision-making phase, fix an unassigned signal, and update the transitive closure to determine all logical consequences of this decision.

The transitive closure computation is similar to matrix multiplication and, hence, easily parallelizable. Parallel computation of transitive closure has been extensively studied and even constant-time algorithms have been proposed [20].

10.2 Transitive Closure Definition

Given a directed graph $G = (V, E)$ with vertex set $V = \{x_1, \ldots, x_n\}$, and edge set E, we wish to find out whether there is a path in G from x_i to x_j for all vertex pairs $x_i, x_j \in V$. The *transitive closure* [1] of G is defined

as the graph $G^* = (V, E^*)$, where

$$E^* = \{(x_i, x_j) : \text{there is a path from vertex } x_i \text{ to vertex } x_j \text{ in } G\}$$

Computing the transitive closure is an important step in many parallel algorithms related to directed graphs [15]. An important application of transitive closure is in deriving all logical conclusions implied by a set of binary relations. It has many other applications.

Finding the transitive closure of a graph G having n vertices requires $O(n^3)$ operations in the worst case [21]. This complexity could be reduced to $O(n^{2.49})$ by exploiting the equivalence between matrix multiplication and transitive closure and using fast matrix multiplication methods. An alternative method for deriving all logical conclusions in *linear expected-time* complexity has also been proposed [11]. There are efficient algorithms for computing the transitive closure of sparse graphs [7].

The computation of transitive closure can easily be accelerated through parallel processing since it belongs to class *NC*, a hierarchy of problems solvable by deterministic algorithms that operate in polylog time using a polynomial-bounded number of processors. If A is the adjacency-matrix of graph G, the transitive closure of G can be obtained by squaring the adjacency matrix $\log_2 n$ times, where n is the number of vertices in graph G. It is known [6] that $n \times n$ matrix multiplication can be performed in $O(\log_2 n)$ time using $O(n^{2.376})$ processors. Thus, transitive closure can be computed in polylog time using the same number of processors. If fewer processors are available, as pointed out in Section 9.5.1, we can use *Brent's scheduling principle* [3] to design an efficient parallel algorithm. Recently, a *constant-time* algorithm has been proposed for computing transitive closure on a processor array with reconfigurable bus system [20].

10.3 Implication Graphs

We represent the digital circuit as a set of binary and ternary relations [5, 13]. As an example, consider the NAND gate shown in Figure 10.1. Since $c = \overline{ab}$, it is easy to see that the equation $F_{NAND} = c \oplus \overline{ab} = 0$ is

FIGURE 10.1: A NAND gate.

satisfied only by those values of a, b and c that satisfy the NAND gate function. Here, $\oplus$ denotes the logical exclusive-OR operation. We will refer to F_{NAND} as the *Boolean false function*. An analogous *Boolean truth function* was used by Larrabee [13]. Using Boolean algebra, we can rewrite F_{NAND} as follows:

$$F_{NAND} = \bar{a}\,\bar{c} + \bar{b}\,\bar{c} + abc \qquad (10.1)$$

where $+$ denotes the logical OR operation. A similar formulation was given in Chapter 5 for representing the NAND gate as an energy function where a, b and c were treated as arithmetic variables. The function F_{NAND} assumes the value 0 only when all the three terms in Equation 10.1 simultaneously become 0. The term $\bar{a}\,\bar{c}$ assumes the value 0 under any one of the following conditions:

1. $a = 1$.

2. $\bar{a} = 1$ and $c = 1$.

3. $c = 1$.

4. $\bar{c} = 1$ and $a = 1$.

Conditions 1 and 2 are equivalent to the binary relation $\bar{a} \Rightarrow c$, where $\Rightarrow$ denotes logical implication. Conditions 3 and 4 are equivalent to the binary relation $\bar{c} \Rightarrow a$. The term $\bar{b}\,\bar{c}$ similarly results in the binary relations $\bar{b} \Rightarrow c$ and $\bar{c} \Rightarrow b$. The term abc does not produce any binary relations. However, if any one of the signals assumes a known value, this term will reduce to a binary relationship. A similar analysis can be performed for all other Boolean gates. The false function for a digital circuit is the logical OR of the false functions for all gates.

It is convenient to represent the binary relations as a directed graph, also called an *implication graph* [2]. This graph for the NAND gate, as shown in Figure 10.2, contains vertices for true and false states of variables. Thus, the vertex set is $\{a, b, c, \bar{a}, \bar{b}, \bar{c}\}$. Equation $\bar{a}\,\bar{c} = 0$ is expressible as binary relations $\bar{a} \Rightarrow c$ and $\bar{c} \Rightarrow a$ that are represented as arcs $(\bar{a}, c)$ and $(\bar{c}, a)$, respectively. Similarly, $\bar{b}\,\bar{c} = 0$ adds the arcs $(\bar{b}, c)$ and $(\bar{c}, b)$ to the implication graph. The transitive closure of this implication graph is shown by its adjacency matrix in Figure 10.3. It can be determined using standard graph-theoretic techniques [7]. Incidentally, in this case, the implication graph also happens to be its own transitive closure.

10.4   A Test Generation Algorithm

We will illustrate our method by an example. Consider the circuit shown in Figure 10.4. We will derive a test for a s-a-0 fault on line c.

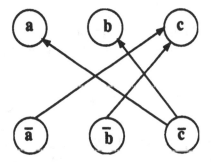

FIGURE 10.2: Implication graph for NAND gate.

	a	$\bar{a}$	b	$\bar{b}$	c	$\bar{c}$
a	1	0	0	0	0	0
$\bar{a}$	0	1	0	0	1	0
b	0	0	1	0	0	0
$\bar{b}$	0	0	0	1	1	0
c	0	0	0	0	1	0
$\bar{c}$	1	0	1	0	0	1

FIGURE 10.3: Transitive closure of the NAND gate.

First, we derive two types of constraints to be represented in the implication graph:

1. Functional constraints that depend upon the function of gates used in the circuit.

2. Structural constraints that depend on the specific interconnection topology of gates and are independent of the gate types used in the circuit.

Functional constraints: Signals that lie on a path from the fault site to a primary output may assume different values in the fault-free and faulty

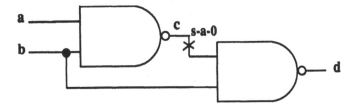

FIGURE 10.4: An example circuit.

circuits. Therefore, additional binary variables are assigned to these signals. In our example, signals c and d lie on the path from the fault site to the primary output and we assign two binary variables c' and d', respectively. Conceptually, this can be visualized as construction of a faulty circuit copy, as shown in Figure 10.5. The Boolean false function for the modified circuit is given by the logical OR of the false functions for the

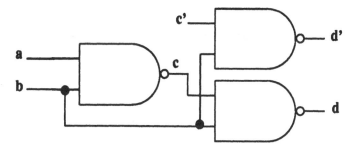

FIGURE 10.5: Modified circuit for fault on c.

gates c, d and d' (we address a gate by the name of its output signal). Furthermore, signals c and c' are constrained to assume the logic values 1 and 0, respectively. All the functional constraints can be determined from the netlist of the circuit in time complexity that is linear in the number of signals.

Structural constraints: We assign a binary variable, called the *path* variable, s_x to every signal x that lies on a path from the fault site to the primary output. This variable assumes the logic value 1 only when the fault on x is observable at the primary output. In our example, signals c and d are assigned the binary variables s_c and s_d. There are two types of structural constraints. Both can be derived from the netlist of the circuit in time complexity that is linear in the number of signals. First, an input fault on a gate is observable only if the gate output fault is observable. Note that the converse need not be true. In our example, if $s_c = 1$, then s_d must assume the value 1. This constraint is expressed as follows: $s_c \bar{s}_d = 0$. For a stem fault to be observable, the fault must propagate through one or more of the branches. This case is not applicable in the present example. Second, if the path variable associated with signal x is true, then signal x must assume different values in the fault-free and faulty circuits. If we denote the faulty circuit value as x', this constraint is expressed as follows: $s_x(xx' + \bar{x}\,\bar{x}') = 0$. Notice that this condition cannot be expressed as binary relations. For the fault to be observable, the path variables associated with the fault site and the primary output are constrained to be 1. Therefore, in our example, $s_c = s_d = 1$.

In summary, the functional constraints are as follows:

$$\bar{a}\,\bar{c} + \bar{b}\,\bar{c} + abc = 0$$
$$\bar{b}\,\bar{d} + \bar{c}\,\bar{d} + bcd = 0$$
$$\bar{b}\,\bar{d'} + \bar{c'}\,\bar{d'} + bc'd' = 0$$

The structural constraints are as follows:

$$s_c\,\bar{s}_d = 0$$
$$s_c(cc' + \bar{c}\,\bar{c'}) = 0$$
$$s_d(dd' + \bar{d}\,\bar{d'}) = 0$$

The logical OR of the functional and structural constraints is the Boolean false function for the circuit with fault. Test generation proceeds as follows:

1. Derive functional and structural constraints for the fault.

2. Determine the transitive closure of constraints expressible as binary relations and identify contradictions, identifications and fixations (derivation of logical conclusions from the transitive closure is described later). If a contradiction is detected, the fault is redundant. If signal values, thus determined, satisfy the false function, we have a test. Otherwise, a partial set of signal values determined thus far may reduce some of the ternary relations to binary relations. We include these relations in the transitive closure and recompute fixations and contradictions. We continue this process until no ternary relation reduces to a binary relation.

3. Make a choice on an unassigned decision variable. This may cause some of the ternary relations to become binary relations.

4. Include the new binary relations in the transitive closure and observe any contradictions, identifications and fixations. If there is a contradiction, we backtrack to Step 3 and make an alternative choice. If the signal assignments satisfy the false function, we have a test. Otherwise, ternary relations that have been reduced to binary relations due to current set of signal assignments are added to the transitive closure and we go to Step 2.

Since we are deriving a test for the s-a-0 fault on signal c, $c = 1$, $c' = 0$, $s_c = s_d = 1$ and our constraint set reduces to:

$$ab = 0$$
$$\bar{b}\bar{d} + db = 0$$
$$\bar{d} + \bar{b}\bar{d} = 0$$
$$dd' + \bar{d}\bar{d} = 0$$

All the terms can be expressed as binary relations. Let R denote the set of these binary relations. The implication graph is shown in Figure 10.6. It

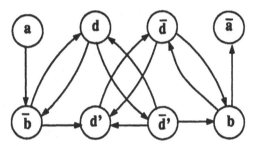

FIGURE 10.6: Implication graph for circuit in Figure 10.5.

contains vertices for variables and their complements. Here, the vertex set is $\{a, b, d, d', \bar{a}, \bar{b}, \bar{d}, \bar{d}'\}$. The constraint $ab = 0$ is expressible as a pair of binary relations $a \Rightarrow \bar{b}$ and $b \Rightarrow \bar{a}$ that are represented as arcs $(a, \bar{b})$ and $(b, \bar{a})$, respectively. Arcs corresponding to the other terms are similarly derived.

The transitive closure of the implication graph is shown in Figure 10.7. This can be computed using standard graph-theoretic techniques [7]. The following types of logical conclusions are implied by the relations in R and they can all be determined from the transitive closure as follows:

1. *Contradiction:* Some variable x must assume both 0 and 1 value. This happens when the transitive closure consists of both arcs $(x, \bar{x})$ and $(\bar{x}, x)$. If this situation occurs before the search process begins or after the search space has been implicitly exhausted, the fault is redundant. If a contradiction occurs during the search process, we must backtrack to the previous choice.

2. *Identification:* Certain pairs of literals must assume the same value. If the transitive closure consists of the arcs (x, y) and (y, x), then literals x and y must assume the same values.

	a	$\bar{a}$	b	$\bar{b}$	d	$\bar{d}$	d'	$\bar{d'}$
a	1	1	1	1	1	1	1	1
$\bar{a}$	0	1	0	0	0	0	0	0
b	0	1	1	0	0	1	1	0
$\bar{b}$	0	1	1	1	1	1	1	1
d	0	1	1	1	1	1	1	1
$\bar{d}$	0	1	1	0	0	1	1	0
d'	0	1	1	0	0	1	1	0
$\bar{d'}$	0	1	1	1	1	1	1	1

FIGURE 10.7: Transitive closure of implication graph in Figure 10.6.

3. *Fixation*: Some variable x can be fixed to the value 0 or 1. If the transitive closure consists of the arc $(x, \bar{x})$ but does not have the arc $(\bar{x}, x)$, then x can be fixed at 0. Similarly, if the transitive closure consists of the arc $(\bar{x}, x)$ but not $(x, \bar{x})$, variable x can be fixed at 1.

The transitive closure in Figure 10.7 contains the arc $(a, \bar{a})$ and, therefore, $a = 0$. Similarly, $b = d' = 1$ and $d = 0$. Since all primary input signals have been determined, we do not need the search phase. The vector $a = 0, b = 1$ is a test for the fault c s-a-0.

10.5 Identifying Necessary Assignments

The techniques of implication, justification, static learning, dynamic learning and sensitization of dominators are used by many test generation algorithms for determining necessary signal assignments and redundancies. All these techniques are implicit in the transitive closure method. We will present simple examples to show that the transitive closure method subsumes these techniques. Also, since transitive closure determines all logical consequences, it is possible to determine redundancies and necessary signal assignments that may not be possible using other techniques. As an example, *logical* dominators are implicitly sensitized. Due to the partial set of signal assignments, it is possible that the fault effect must propagate through some gate that is not a dominator of the fault. Such gates are labeled as logical dominators. Section 10.5.3 describes the implicit sensitization of dominators. Furthermore, it is possible to detect difficult redundancies that may not be detectable using other techniques. Section 10.5.4 describes the application of transitive closure in redundancy identification.

10.5.1 Implicit Implication and Justification

Consider the circuit shown in Figure 10.8. The Boolean false function for this circuit, which is the logical OR of the false functions for gates d, e and f, consists of the following two-variable terms: $\bar{a}\,\bar{d}, \bar{b}\,\bar{d}, \bar{b}\,\bar{e}, \bar{c}\,\bar{e}, \bar{d}\,\bar{f}$ and $\bar{e}\,\bar{f}$. The binary relations arising from these terms are represented in the

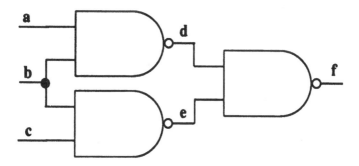

FIGURE 10.8: An example circuit.

implication graph shown in Figure 10.9. This graph also happens to be its own transitive closure. The adjacency matrix of the transitive closure is

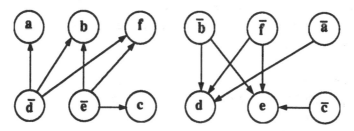

FIGURE 10.9: Implication graph of circuit in Figure 10.8.

shown in Figure 10.10.

Suppose $d = 0$. Forward implication will result in $f = 1$ and justification will result in $a = b = 1$. This is implicitly determined in the transitive closure method:

1. Include the relation $d = 0$ into the implication graph of Figure 10.9 by adding the arc $(d, \bar{d})$. The updated implication graph is shown in Figure 10.11.

2. Construct the transitive closure of the updated implication graph. The adjacency matrix of the updated transitive closure is shown in Figure 10.12. We can directly derive the updated transitive closure from the transitive closure shown in Figure 10.10 as follows. The

	a	$\bar{a}$	b	$\bar{b}$	c	$\bar{c}$	d	$\bar{d}$	e	$\bar{e}$	f	$\bar{f}$
a	1	0	0	0	0	0	0	0	0	0	0	0
$\bar{a}$	0	1	0	0	0	0	1	0	0	0	0	0
b	0	0	1	0	0	0	0	0	0	0	0	0
$\bar{b}$	0	0	0	1	0	0	1	0	1	0	0	0
c	0	0	0	0	1	0	0	0	0	0	0	0
$\bar{c}$	0	0	0	0	0	1	0	0	1	0	0	0
d	0	0	0	0	0	0	1	0	0	0	0	0
$\bar{d}$	1	0	1	0	0	0	0	1	0	0	1	0
e	0	0	0	0	0	0	0	0	1	0	0	0
$\bar{e}$	0	0	1	0	1	0	0	0	0	1	1	0
f	0	0	0	0	0	0	0	0	0	0	1	0
$\bar{f}$	0	0	0	0	0	0	1	0	1	0	0	1

FIGURE 10.10: Transitive closure of implication graph in Figure 10.9.

addition of the arc $(d, \bar{d})$ will allow all ancestors of d to reach all descendants of $\bar{d}$. The ancestors of d, *i.e.*, vertices $\bar{a}$, $\bar{b}$ and $\bar{f}$, are the vertices with a 1 in column d in Figure 10.10 and the descendants of $\bar{d}$, *i.e.*, vertices a, b and f, are the vertices with a 1 in row $\bar{d}$. For example, vertex $\bar{b}$ (an ancestor of d) that in Figure 10.10 could only reach d and e, can now also reach vertices a, b, $\bar{d}$ and f (see Figure 10.12).

3. Derive logical conclusions from the updated transitive closure. In Figure 10.12, there is an arc $(\bar{a}, a)$ but there is no arc $(a, \bar{a})$. Thus, $a = 1$ is the only value that can satisfy the arc $(\bar{a}, a)$ (note that this arc can only result from the relation $\bar{a} = 0$). Similarly, we can conclude that $b = f = 1$.

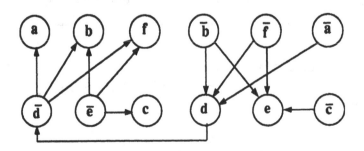

FIGURE 10.11: Implication graph of circuit in Figure 10.8 with $d = 0$.

	a	$\bar{a}$	b	$\bar{b}$	c	$\bar{c}$	d	$\bar{d}$	e	$\bar{e}$	f	$\bar{f}$
a	1	0	0	0	0	0	0	0	0	0	0	0
$\bar{a}$	1	1	1	0	0	0	1	1	0	0	1	0
b	0	0	1	0	0	0	0	0	0	0	0	0
$\bar{b}$	1	0	1	1	0	0	1	1	1	0	1	0
c	0	0	0	0	1	0	0	0	0	0	0	0
$\bar{c}$	0	0	0	0	0	1	0	0	1	0	0	0
d	1	0	1	0	0	0	1	1	0	0	1	0
$\bar{d}$	1	0	1	0	0	0	0	1	0	0	1	0
e	0	0	0	0	0	0	0	0	1	0	0	0
$\bar{e}$	0	0	1	0	1	0	0	0	0	1	1	0
f	0	0	0	0	0	0	0	0	0	0	1	0
$\bar{f}$	1	0	1	0	0	0	1	1	1	0	1	1

FIGURE 10.12: Transitive closure of implication graph in Figure 10.11.

10.5.2 Transitive Closure Does More Than Implication and Justification

Consider the case when signal f in the circuit of Figure 10.8 is assigned the value 1. This results in the addition of the arc $(\bar{f}, f)$ to the implication graph leading to Figure 10.13. The transitive closure method will fix signal b to logic value 1, as explained below. It is important to note that, given $f = 1$, conventional implication and justification procedures will not be able to conclude that signal b should be fixed at 1.

If $f = 1$, the three-variable term def in the false function of the gate f reduces to de. Inclusion of the relations $\bar{f} = 0$ and $de = 0$ will result in the addition of the following arcs to the implication graph of Fig-

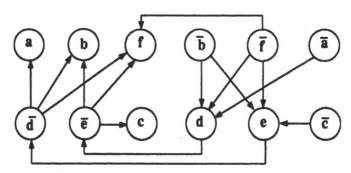

FIGURE 10.13: Implication graph of circuit in Figure 10.8 with $f = 1$.

ure 10.9: $(\bar{f}, f), (d, \bar{e})$ and $(e, \bar{d})$. The updated implication graph is shown in Figure 10.13 and the corresponding transitive closure adjacency matrix is given in Figure 10.14. As explained earlier, we can also construct the updated transitive closure by incrementally changing the transitive closure shown in Figure 10.10.

	a	$\bar{a}$	b	$\bar{b}$	c	$\bar{c}$	d	$\bar{d}$	e	$\bar{e}$	f	$\bar{f}$
a	1	0	0	0	0	0	0	0	0	0	0	0
$\bar{a}$	0	1	1	0	1	0	1	0	0	1	1	0
b	0	0	1	0	0	0	0	0	0	0	0	0
$\bar{b}$	1	0	1	1	1	0	1	1	1	1	1	0
c	0	0	0	0	1	0	0	0	0	0	0	0
$\bar{c}$	1	0	1	0	0	1	0	1	1	0	1	0
d	0	0	1	0	1	0	1	0	0	1	1	0
$\bar{d}$	1	0	1	0	0	0	0	1	0	0	1	0
e	1	0	1	0	0	0	0	1	1	0	1	0
$\bar{e}$	0	0	1	0	1	0	0	0	0	1	1	0
f	0	0	0	0	0	0	0	0	0	0	1	0
$\bar{f}$	1	0	1	0	1	0	1	1	1	1	1	1

FIGURE 10.14: Transitive closure of implication graph in Figure 10.13.

Notice that the updated transitive closure (Figure 10.14) has a 1 in row $\bar{b}$ and column b. Thus, $\bar{b}$ implies b. But b does not imply $\bar{b}$. The arc $(\bar{b}, b)$, therefore, indicates the relation $\bar{b} = 0$. Hence, b must be fixed at 1 as a consequence of originally fixing f to 1.

10.5.3 Implicit Sensitization of Dominators

Dominators are gates through which the fault must propagate in order to be observable [12]. We refer to such gates as *physical* dominators. As an example, consider the fault b s-a-0 in the circuit of Figure 10.8. It is easy to see that the fault must propagate through gate f to be observable at the output of the circuit and thus f is identified as a physical dominator. When signal a is assigned the value 0, the fault must propagate through gate e to reach an output. Gate e is now a logical dominator.

The transitive closure method implicitly identifies and sensitizes many physical and logical dominators. We will illustrate these features by an example. The physical dominators of fault b s-a-0 are determined purely from the structural constraints, as expressed by these path variables: $s_a \bar{s}_d = 0$, $s_c \bar{s}_e = 0$, $s_d \bar{s}_f = 0$, $s_e \bar{s}_f = 0$ and $s_b \bar{s}_d \bar{s}_e = 0$, The last constraint ensures that the stem fault on b propagates to either signal d or e or both in order to be observable at the primary output. Initially, s_b is set to 1 since we assume that the fault on stem b is observable. This will reduce the ternary relation $s_b \, \overline{s}_d \, \overline{s}_e = 0$ to the binary relation $\overline{s}_d \, \overline{s}_e = 0$. It can easily be verified that the transitive closure of these binary relations fixes s_f to 1. Therefore, gate f is a dominator. Note that f will be identified as a dominator of the fault even if this circuit was embedded in a larger circuit.

As an example of identification and sensitization of logical dominators, consider the case when signal a is set to logic value 0. Since the fault on stem b cannot propagate to gate d, gate e becomes the logical dominator. Furthermore, signal c must be set to 1 to sensitize the logical dominator e. The identification of e as the logical dominator and the assignment $c = 1$ is implicitly determined by the transitive closure method, even in the case when the circuit is part of a larger circuit and there are multiple paths from signal f to the primary output. To see this, we construct the Boolean false function for the circuit with fault. Conceptually, we can visualize the circuit with fault as the modified circuit shown in Figure 10.15. By including

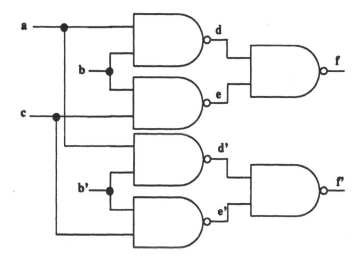

FIGURE 10.15: Modified circuit for fault b s-a-0 in Figure 10.8.

the functional and structural constraints in the Boolean false function of the faulty circuit, and by computing the transitive closure of the binary relations, we see that $s_e = 1$ and $c = 1$.

10.5.4 Redundancy Identification

Consider the circuit shown in Figure 10.16. By simulating the circuit for all possible combinations of a and b, it can be verified that the fault k s-a-1 is indeed redundant. Such redundancies are, in general, more difficult to detect than others. Prior techniques like implication, justification, static and dynamic learning and others are unable to detect this redundancy. We

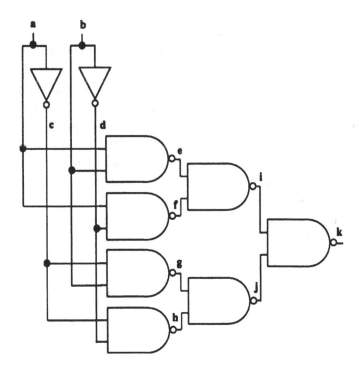

FIGURE 10.16: A circuit with redundant fault k s-a-1.

will show that the transitive closure method implicitly identifies this redundancy. It is important to note that our method will identify such a redundancy even when this circuit is embedded in a much larger circuit. Furthermore, it is possible to remove redundant logic using the transitive closure.

Since the fault is on the primary output, we simply have to justify the value 0 on signal k. The binary relations obtained from the Boolean false function of the circuit are listed in Table 10.1. The Boolean false function for gate k is $\bar{i} + \bar{j}$ since $k = 0$. Therefore, gate k contributes the relations $\bar{i} = 0$ and $\bar{j} = 0$ to the Boolean false function of the entire circuit. Binary relations contributed by other gates can be similarly derived.

If we represent these relations as an implication graph and compute the transitive closure, both arcs $(a, \bar{a})$ and $(\bar{a}, a)$ will be present in the transitive closure. This indicates a contradiction and, therefore, no assignment of signal values will satisfy the Boolean false function for the circuit with $k = 0$. We conclude that the fault is redundant.

TABLE 10.1: Binary relations for circuit in Figure 10.16 ($k = 0$).

Gate	Binary relations
c	$ac = 0$ and $\bar{a}\,\bar{c} = 0$
d	$bd = 0$ and $\bar{b}\,\bar{d} = 0$
e	$\bar{a}\,\bar{e} = 0$ and $\bar{b}\,\bar{e} = 0$
f	$\bar{a}\,\bar{f} = 0$ and $\bar{d}\,\bar{f} = 0$
g	$\bar{b}\,\bar{g} = 0$ and $\bar{c}\,\bar{g} = 0$
h	$\bar{c}\,\bar{h} = 0$ and $\bar{d}\,\bar{h} = 0$
i	$\bar{e}\,\bar{\imath} = 0$ and $\bar{f}\,\bar{\imath} = 0$
j	$\bar{g}\,\bar{\jmath} = 0$ and $\bar{h}\,\bar{\jmath} = 0$
k	$\bar{\imath} = 0$ and $\bar{\jmath} = 0$

10.6 Summary

We have applied the graph theoretic concept of transitive closure to test generation. There are several advantages of this approach. First, the transitive closure is a single tool that replaces the entire *bag of tricks* used to speed up branch-and-bound search. Second, our technique determines *all* logical consequences based on pairwise signal relationships for a partial set of signal assignments and provides a good framework for reasoning about signal relationships in the circuit. Third, transitive closure computation is easily parallelizable. Therefore, it can be effectively used to identify necessary assignments in parallel test generation methods. Also, in conventional test generation methods, implication and justification are carried out in a series of steps. Transitive closure integrates many steps in a global sense. Thus, conflicts are quickly discovered. We have indicated the application of transitive closure in redundancy identification.

While we have presented a new algorithm, the transitive closure can be included in any existing test generation algorithm like PODEM, FAN, TOPS or SOCRATES, to speed up the search for a test by quickly identifying necessary signal assignments.

References

[1] A. V. Aho, J. E. Hopcroft, and J. D. Ullman. *The Design and Analysis of*

Computer Algorithms. Addison-Wesley Publishing Company, Reading, MA, 1974.

[2] B. Aspvall, M. F. Plass, and R. E. Tarjan. A Linear-Time Algorithm for Testing the Truth of Certain Quantified Boolean Formulas. *Information Processing Letters*, 3(8):121–123, March 1979.

[3] R. P. Brent. The Parallel Evaluation of General Arithmetic Expressions. *Journal of the ACM*, 21(2):201–206, April 1974.

[4] S. T. Chakradhar and V. D. Agrawal. A Transitive Closure Based Algorithm for Test Generation. In *Proceedings of the 28th ACM/IEEE Design Automation Conference*, June 1991.

[5] S. T. Chakradhar, V. D. Agrawal, and M. L. Bushnell. Neural Net and Boolean Satisfiability Models of Logic Circuits. *IEEE Design and Test of Computers*, 7(5):54–57, October 1990.

[6] D. Coppersmith and S. Winograd. Matrix Multiplication via Arithmetic Progressions. In *Proceedings of the 19th Annual ACM Symposium on Theory of Computing*, pages 1–6, May 1987.

[7] T. H. Cormen, C. E. Leiserson, and R. L. Rivest. *Introduction to Algorithms.* McGraw Hill, New York, 1990.

[8] H. Fujiwara and T. Shimono. On the Acceleration of Test Generation Algorithms. *IEEE Transactions on Computers*, C-32(12):1137–1144, December 1983.

[9] J. Giraldi and M. L. Bushnell. EST: The New Frontier in Automatic Test-Pattern Generation. In *Proceedings of the 27th ACM/IEEE Design Automation Conference*, pages 667–672, June 1990.

[10] P. Goel. An Implicit Enumeration Algorithm to Generate Tests for Combinational Logic Circuits. *IEEE Transactions on Computers*, C-30(3):215–222, March 1981.

[11] P. Hansen, B. Jaumard, and M. Minoux. A Linear Expected-Time Algorithm for Deriving All Logical Conclusions Implied by a Set of Boolean Inequalities. *Mathematical Programming*, 34(2):223–231, March 1986.

[12] T. Kirkland and M. R. Mercer. A Topological Search Algorithm For ATPG. In *Proceedings of the 24th ACM/IEEE Design Automation Conference*, pages 502–508, June 1987.

[13] T. Larrabee. Efficient Generation of Test Patterns Using Boolean Difference. In *Proceedings of the IEEE International Test Conference*, pages 795–801, August 1989.

[14] S. Patil and P. Banerjee. A Parallel Branch and Bound Algorithm for Test Generation. *IEEE Transactions on Computer-Aided Design*, 9(3):313–322, March 1990.

[15] S. R. Pawagi, P. S. Gopalakrishnan, and I. V. Ramakrishnan. Computing Dominators in Parallel. *Information Processing Letters*, 24(4):217–221, 1987.

[16] J. Rajski and H. Cox. A Method to Calculate Necessary Assignments in Algorithmic Test Pattern Generation. In *Proceedings of the IEEE International Test Conference*, pages 25–34, September 1990.

[17] M. H. Schulz and E. Auth. Improved Deterministic Test Pattern Generation with Applications to Redundancy Identification. *IEEE Transactions on Computer-Aided Design*, 8(7):811–816, July 1989.

[18] M. H. Schulz, E. Trischler, and T. M. Sarfert. SOCRATES: A Highly Efficient Automatic Test Pattern Generation System. *IEEE Transactions on Computer-Aided Design*, 7(1):126–136, January 1988.

[19] V. Sivaramakrishnan, S. C. Seth, and P. Agrawal. Parallel Test Pattern Generation using Boolean Satisfiability. In *Proceedings of the 4th CSI/IEEE International Symposium on VLSI Design, New Delhi*, pages 69–74, January 1991.

[20] B. F. Wang and G. H. Chen. Constant Time Algorithms for the Transitive Closure and Some Related Graph Problems on Processor Arrays with Reconfigurable Bus System. *IEEE Transactions on Parallel and Distributed Systems*, 1(4):500–507, October 1990.

[21] S. Warshall. A Theorem on Boolean Matrices. *Journal of the ACM*, 9(1):11–12, January 1962.

Chapter 11

POLYNOMIAL-TIME TESTABILITY

> *"In the theory of computation, there is general agreement that a problem should be considered intractable if it cannot be solved in less than exponential time. ... suppose that we have three algorithms whose time complexities are n, n^3 and 2^n. Assuming that time complexity expresses execution time in microseconds, the n, n^3 and 2^n algorithms can solve a problem of size 10 instantly in 0.00001, 0.001 and 0.001 seconds, respectively. ... to solve a problem of only size 60, the 2^n algorithm requires 366 centuries whereas the n and n^3 algorithms require only 0.00006 and 0.216 seconds, respectively."*
>
> – H. Fujiwara in *Logic Testing and Design for Testability*, MIT Press (1985)

The problem of detecting a fault in a general combinational circuit is NP-complete [5] and it is unlikely that a polynomial time algorithm exists for solving it. The non-polynomial time complexity here refers to the *worst-case* effort of test generation in a circuit. Consequently, it is of interest to identify circuits for which a polynomial time fault detection algorithm exists. The number of primary inputs and the number of signals in the digital circuit are generally considered as the the *input size* for the fault detection problem.

123

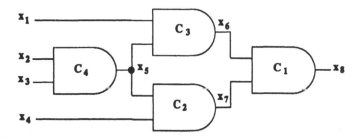

FIGURE 11.1: A $(2, 2)$-circuit.

Very little work has been done on identifying easily-testable circuits. Such work is important for two reasons. First, solving special instances of the test generation problem may provide a better insight into the solution methods for the general problem. Second, combinational circuits can be synthesized so that they are easily testable and easy for redundancy identification. In this chapter, we present results that might take us a step closer to the goal of integrating design and test.

11.1 Background

An earlier result in this area is due to Fujiwara [4] who defined $K-bounded$ circuits. He assumed that logic blocks with no more than K input lines are so connected that *no* reconvergent fanout exists among these blocks. Essentially, the undirected graph G, with a vertex for each block and an edge between a pair of vertices whenever the corresponding blocks are connected, is a tree. He showed that for such circuits, the single stuck-at fault detection problem is polynomial time solvable.

In this chapter, we define a larger class of circuits, called $(k, K)-circuits$, and present a polynomial time algorithm to detect any single or multiple stuck-at fault [3]. The logic blocks in a $(k, K)-$circuit are still $K-$bounded, i.e., these blocks have no more than K input lines. However, we permit certain cycles in the associated graph G. Specifically, G is a *partial $k-$tree* [6]. Any subgraph of a $k-tree$ (a graph with no cliques of size greater than $k + 1$) is a partial $k-$tree. Thus, the main difference between a $K-$bounded circuit and a $(k, K)-$circuit is that the latter can have certain reconvergent fanouts among the blocks while the former has none.

Any combinational circuit can be considered an interconnection of several sub-circuits called *blocks*. A block can have several inputs and outputs. For example, an inverter is a block with one input and one output and a two-input AND gate is a block with two inputs and one output. In general, a

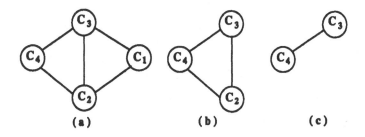

FIGURE 11.2: Graphs G_{π_1} (a), G_{π_2} (b) and G_{π_3} (c).

block may have several gates. The *union* ($\cup$) of two blocks is defined as the set of gates in both blocks and their *intersection* ($\cap$) is the set of gates common to both blocks. We define an arbitrary combinational circuit C as an interconnection of blocks $C_1, \ldots, C_t$ such that $C = C_1 \cup C_2 \cup \ldots \cup C_t$ and $C_i \cap C_j = 0$ where $i \neq j$, $1 \leq i \leq t$ and $1 \leq j \leq t$. We will refer to the set $\pi_1 = \{C_1, C_2, \ldots, C_t\}$ as a *partition* of C. A circuit can be partitioned in several ways. With a partition π_1, we associate an undirected graph $G_{\pi_1}(V, \mathcal{E})$ having a vertex set V and an edge set $\mathcal{E}$. The graph has a vertex for every block C_i in the partition and there is an edge from vertex i to vertex j if the corresponding blocks have at least one signal in common. For example, the circuit in Figure 11.1 can be partitioned into four blocks C_1, C_2, C_3 and C_4. Each block has one gate. Furthermore, $\{x_6, x_7, x_8\}$ is the set of signals in block C_1, $\{x_1, x_5, x_6\}$ is the set of signals in C_3 and so on. The corresponding graph G_{π_1} is shown in Figure 11.2(a). Notice that there is an edge between vertex C_3 and C_2 since they share the signal x_5. We will later discuss the other graphs shown in Figure 11.2.

11.1.1 Fujiwara's Result

A combinational circuit C is K–bounded [4] if it can be partitioned into blocks such that:

1. C_i ($1 \leq i \leq t$) has at most K inputs and

2. G_{π_1} has no cycles.

For example, the circuit in Figure 11.3 is 2–bounded since it can be partitioned into three blocks c, g and h, each with at most two inputs. It is easy to see that the corresponding graph does not have a cycle. However, the circuit in Figure 11.1 is *not* 2-bounded since the corresponding graph, shown in Figure 11.2(a), has cycles, e.g., vertices C_2, C_3 and C_4 form a cycle.

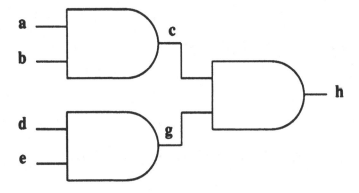

FIGURE 11.3: Example logic circuit.

The single stuck-at fault detection problem in K–bounded circuits, for a fixed K, is solvable in $O(g)$ time, where g is the number of signals in the circuit [4].

11.1.2 Contribution of the Present Work

We relax the restriction on the graph G_{π_1} by allowing certain types of cycles. Our new class of circuits, called the (k, K)–*circuits*, can be partitioned into blocks such that

1. C_i ($1 \leq i \leq t$) has at most K inputs and

2. G_{π_1} is a partial k–tree.

A partial k–tree graph is a subgraph of a k–*tree* [6]. A k–tree is a graph that can be reduced to the k–complete graph (i.e., a fully-connected graph on k vertices) by a sequence of removals of a degree k vertex with completely connected neighbors [6].

A constructive definition of a k–tree is given in Section 11.2. A graph G_{π_1} with no cycles is a special case of a partial k–tree (partial 1–tree) and thus, all K–bounded circuits are $(1, K)$–circuits.

Note that in K–bounded circuits, the parameter K imposes a limit on the number of inputs to a block and these blocks are assumed to be connected such that the associated graph has no cycles. In (k, K)–circuits, all blocks still have at most K inputs but the parameter k gives greater freedom in interconnecting the blocks. Observe that we are imposing a restriction on gate interconnections rather than on the function of gates comprising the circuit. Fixed values of k and K specify a class of circuits. By varying the two parameters, a whole family of circuits can be defined.

We show that the single and multiple stuck-at fault detection problem is solvable in polynomial time for (k, K)–circuits. Our proposed algorithm for doing this is radically different from the traditional fault detection algorithms [1]. Using the neural net model of logic circuits, proposed in Chapter 5, we formulate fault detection as an energy minimization problem and show that a minimizing point (global minimum) of the energy function can be determined in time complexity that is a polynomial in the size of the circuit. In this chapter, we use the terms *minimizing point* and *global minimum* interchangeably. Section 11.2 reviews partial k–tree graphs and discusses notation and terminology used in the remainder of this chapter. In Section 11.3, we present a polynomial time algorithm for detecting any single or multiple fault in a (k, K)–circuit.

11.2 Notation and Terminology

Consider a combinational circuit C consisting of an interconnection of inverters and 2-input Boolean gates. Let the total number of signals (primary inputs and gate outputs) be g. Let $\pi_1 = \{C_1, C_2, \ldots, C_t\}$ be the partition associated with the circuit C (see Table 11.1 for a list of symbols). We will refer to C_i as the ith block of C. A block consists of one or more gates. An input or output signal of a block is called a *port signal* and a signal that is not an input or output signal of a block is called an *internal signal*. Let g_i be the number of signals in the ith block. For example, in Figure 11.1, x_1, x_5 and x_6 are the port signals of block C_3. None of the blocks have internal signals.

Consider a graph $G(V, \mathcal{E})$, where V is the vertex set and $\mathcal{E}$ the edge set of G. $G \setminus v$ will denote the subgraph induced by the set $V \setminus v$ (i.e., all vertices except v) where v is a vertex in the graph. A fully connected graph on k vertices is called a k–*complete* graph [2]. A k–tree [6] is a graph that can be reduced to the k–complete graph by a sequence of removal of degree k vertices with completely connected neighbors. This vertex removal sequence is called the k–*perfect elimination scheme* of the k–tree. For example, the graph in Figure 11.2(a) is a 2-tree because it can be reduced to a 2-complete graph as follows (see Figures 11.2(b) and 11.2(c)). Vertex C_1 is a degree 2 vertex with completely connected neighborhood consisting of vertices C_2 and C_3. If we remove C_1 then C_2 becomes a degree 2 vertex with completely connected neighborhood comprising of vertices C_3 and C_4. Removal of vertex C_2 leaves us with a 2-complete graph consisting of the single edge (C_3, C_4). The sequence of removing C_1 and C_2 is a perfect elimination scheme for this graph. In general, a graph can have several perfect elimination schemes. For example, the sequence of removing C_4 and

TABLE 11.1: List of symbols.

g	number of signals in circuit C.
t	number of blocks in circuit C.
C_i	ith block in circuit C.
g_i	number of signals in block C_i.
π_i	the partition of circuit C minus the first $i-1$ blocks, i.e., the set $\{C_i, C_{i+1}, \ldots, C_t\}$.
G_{π_i}	undirected graph corresponding to partition π_i.
N_i	set of neighbors of block C_i in graph G_{π_i}.
J_i	set of variables of all blocks in N_i except the variables in blocks $C_1, C_2, \ldots, C_{i-1}$.

C_1 is another perfect elimination scheme for the graph in Figure 11.2(a). The class of k–trees is recursively defined as follows [6]:

1. The complete graph with k vertices is a k–tree.

2. A k–tree with $n + 1$ vertices ($n \geq k$) can be constructed from a k–tree with n vertices by adding a vertex adjacent to all vertices of one of its k–vertex complete subgraphs, and only to these vertices.

For example, the graph in Figure 11.2(a) can be constructed from the 2-complete graph consisting of the edge (C_3, C_4) as follows. Add vertex C_2 adjacent to all vertices in the 2-complete graph of vertices C_3 and C_4. This results in the edges (C_2, C_4) and (C_2, C_3). Similarly, add vertex C_1 adjacent to all vertices in the 2-complete subgraph consisting of the edge (C_2, C_3). This results in the edges (C_2, C_1) and (C_3, C_1).

A partial k–tree is a subgraph of a k–*tree*. For example, consider the 2-tree graph in Figure 11.2(a). The subgraph with vertices C_1, C_2 and C_3 is a partial 2-tree. So is the subgraph with vertices C_1, C_3 and C_4. Obviously, since the entire graph can be treated as a subgraph, the entire graph is a partial 2-tree. The perfect elimination scheme for the partial k–tree is the same as the elimination sequence for the corresponding k–tree.

11.3 A Polynomial Time Algorithm

We will first consider the problem of detecting a primary output fault in a (k, K)–circuit and then show how an arbitrary single or multiple fault in a (k, K)–circuit can be detected in polynomial time ($O(g^2)$).

11.3.1 Primary Output Fault

Consider a single-output circuit C with a s-a-0(1) fault on the primary output. The test must simply control the output to 1(0). It is, therefore, unnecessary to create a faulty circuit copy. We constrain the primary output signal to assume the value 1(0). Let E_i be the energy function associated with the ith block. If the block has more than one gate, then E_i is the sum of the individual gate energy functions. The energy function for the entire circuit is $E_{CKT} = E_1 + E_2 + ... + E_t$. We further simplify E_{CKT} by fixing the required logic value on the primary output signal. Now, any solution of $E_{CKT} = 0$ is a consistent set of signals for the entire circuit and, hence, corresponds to a test for the fault. If no solution is possible, then the fault is redundant.

To show that a minimizing point of E_{CKT}, and hence, a test for the fault, can be obtained in polynomial time, we will need the following lemmas.

Lemma 11.1: *Let C be a (k, K)–circuit with a partition $\pi_1 = \{C_1, C_2, ..., C_t\}$ and let $(C_1, C_2, ..., C_t)$ be the k–perfect elimination scheme of the graph G_{π_1}. Let $E_{1,t} = E_{CKT}$ be the energy function for C. The minimization $E_{1,t}$ can be reduced in $O(g)$ time to the minimization of a function $E_{2,t}$ that depends on variables in blocks $C_2, C_3, ..., C_t$ but not on the port variables of block C_1. Also, the function $E_{2,t}$ has at most a constant number (2^{kK}) of terms more than the function $E_{1,t}$.*

Proof: In order to construct the function $E_{2,t}$ from $E_{1,t}$, we explicitly write the terms of $E_{1,t}$ corresponding to the variables in C_i ($1 \leq i \leq t - k$) that are excluded from $C_1, C_2, ..., C_{i-1}$. We also construct a list of terms that involve variables in blocks C_i, $t - k + 1 \leq i \leq t$, but not in the first $t - k$ blocks. The function $E_{1,t}$ has $O(g)$ terms since each gate contributes at most six terms (see Table 5.1). Our representation can be obtained in time $O(g)$ by scanning through the terms of $E_{1,t}$.

We can rewrite the function $E_{1,t}$ as:

$$E_{1,t} = f_1 + h_1$$

where the function h_1 contains all terms that do not involve variables in block C_1 and f_1 contains only terms involving variables in block C_1. Let $N_1 = \{i : C_i \text{ is connected to } C_1\}$ be the set of neighbors of block C_1 in G_{π_1}. Since G_{π_1} is a partial k–tree, block C_1 is connected at most to k

other blocks. Therefore, f_1 has at most $\sum_{i \in N_1} g_i$ variables other than the variables in block C_1. Let J_1 be the set of these $\sum_{i \in N_1} g_i$ variables. Note that in any consistent labeling of the circuit C, all blocks C_i $(1 \leq i \leq t)$ are also consistently labeled. Although a block has g_i variables, there are at most only a constant number (2^K) of consistent labelings of the signals in block C_i, because the block has at most K inputs. Similarly, there can be at most 2^{kK} consistent labelings of the blocks $\{C_i : i \in N_1\}$. Therefore, the $\sum_{i \in N_1} g_i$ variables in set J_1 can assume at most a constant number (2^{kK}) of combinations of Boolean values. For a given vector α of values for the variables in the set J_1, we can evaluate f_1 for all 2^K combinations of values of the variables in block C_1. Let $f_1(\alpha)$ denote the minimum value the function f_1 attains for the vector α.

Clearly, there exists a minimizing point of function $E_{1,t}$, say (β^*, α^*), where β^* is a vector of values for the variables in block C_1 and α^* is a vector of values for variables in the set J_1, such that the minimum value of f_1 for the given vector α^* is attained at β^*. Also, β^* is a *non-zero vector* (i.e., a vector with at least one non-zero component) *iff* $f_1(\alpha^*) < 0$. This leads us to define a function ψ_1 that depends only on the $\sum_{i \in N_1} g_i$ variables in set J_1, such that

$$\psi_1 \;=\; \sum_{\alpha \in M_1} \min(0, f_1(\alpha)) \times T_\alpha \qquad (11.1)$$

where M_1 is the set of a constant number (2^{kK}) of consistent labelings (vectors) of the blocks $\{C_i : i \in N_1\}$ and T_α is the product of literals corresponding to vector α (i.e., $T_\alpha = \prod_{l \in J_1} l_\alpha$ where $l_\alpha = l$ if variable l is 1 in vector α and $l_\alpha = \bar{l}$ if $l = 0$). We evaluate f_1 at all combinations of values of variables in J_1 and $\min(0, f_1(\alpha))$ is the minimum value the function f_1 attains, given a particular vector α of values for the variables in J_1. Therefore, functions f_1 and ψ_1 have the same minimum value.

We now define a new energy function:

$$E_{2,t} \;=\; \psi_1 + h_1 \qquad (11.2)$$

We have thus reduced the problem of minimization of the original function $E_{1,t}$ that depends on all the blocks C_i $(1 \leq i \leq t)$ to the minimization of $E_{2,t}$, that only depends on blocks C_i $(2 \leq i \leq t)$. Since, block C_1 is eliminated, the partition corresponding to $E_{2,t}$ is $\pi_2 = \{C_2, C_3, \ldots, C_t\}$. A minimizing point $(\beta^* \alpha^*)$ of $E_{1,t}$ can then easily be traced back from any minimizing point α^* of $E_{2,t}$ as explained later. Also, since ψ_1 can have at most a constant number (2^{kK}) of terms, $E_{2,t}$ can have at most a

TABLE 11.2: Terms of $E_{1,4}$.

Block	Terms
C_1	$2x_6x_7 + x_6(6 - 4x_1 - 4x_5) + x_7(6 - 4x_4 - 4x_5)$
C_2	$2x_4x_5 + x_5(6 + 2x_1 - 4x_2 - 4x_3)$
C_3, C_4	$2x_2x_3$

constant number of terms more than $E_{1,t}$ and the representation of $E_{2,t}$ can be obtained in $O(g)$ time. ∎

Example 11.1: Consider the s-a-1 fault on signal x_8 in the circuit of Figure 11.1. We construct the energy function for the circuit and find a minimizing point by recursively eliminating the variables of blocks in the following order: C_1, C_2, C_3, C_4. Here, we illustrate the elimination of variables of block C_1 from the energy function of the circuit. The elimination of other blocks is discussed later. The energy function for the circuit is[1]:

$$
\begin{aligned}
E_{CKT} &= E_1 + E_2 + E_3 + E_4 \\
&= -4x_8(x_6 + x_7) + 2x_6x_7 + 6x_8 \\
&\quad -4x_7(x_4 + x_5) + 2x_4x_5 + 6x_7 \\
&\quad -4x_6(x_1 + x_5) + 2x_1x_5 + 6x_6 \\
&\quad -4x_5(x_2 + x_3) + 2x_2x_3 + 6x_5
\end{aligned}
$$

Since signal x_8 is constrained to assume the value 0, we can simplify E_{CKT} by substituting $x_8 = 0$. Let $E_{1,4}$ be the simplified E_{CKT}. The first row in Table 11.2 gives the terms of $E_{1,4}$ that involve variables in block C_1 (i.e., x_6, x_7 and x_8). The second row gives the terms involving the variables of block C_2 that are not in block C_1 (i.e., x_4 and x_5). The last row gives the terms that have only those variables of the last two blocks C_3 and C_4 that are not in blocks C_1 and C_2. The first row in Table 11.2 is the function f_1. The terms in second and third row comprise the function h_1. $N_1 = \{C_2, C_3\}$ is the set of neighbors of block C_1. The set of variables in f_1 that are not in block C_1 is $J_1 = \{x_1, x_4, x_5\}$. Table 11.3 shows the minimum value of f_1 for each combination of values of the variables in set J_1. For example, when $x_1 = 0$, $x_4 = 0$ and $x_5 = 0$, $f_1 = 2x_6x_7 + 6(x_6 + x_7)$ and it assumes a minimum value 0 when $x_6 = x_7 = 0$. Only the fourth, sixth and eighth rows of min f_1 contribute to ψ_1. We multiply the min f_1 value by the corresponding x_1, x_4 and x_5 literals. From Equation 11.1,

---

[1] As in Chapter 9, here also we use signal name x_i to denote the activation value V_{x_i} of the neuron that models x_i.

TABLE 11.3: Possible values of f_1.

x_1	x_4	x_5	f_1	min f_1
0	0	0	$2x_6x_7 + 6(x_6 + x_7)$	0
0	0	1	$2x_6x_7 + 2(x_6 + x_7)$	0
0	1	0	$2x_6x_7 + 6x_6 + 2x_7$	0
0	1	1	$2x_6x_7 + 2x_6 - 2x_7$	-2
1	0	0	$2x_6x_7 + 2x_6 + 6x_7$	0
1	0	1	$2x_6x_7 - 2x_6 + 2x_7$	-2
1	1	0	$2x_6x_7 + 2x_6 + 2x_7$	0
1	1	1	$2x_6x_7 - 2x_6 - 2x_7$	-2

TABLE 11.4: Terms of $E_{2,4}$.

Block	Terms
C_2	$2x_4x_5(1 - \overline{x_1}) + x_5(6 - 4x_2 - 4x_3)$
C_3, C_4	$2x_2x_3$

$$\begin{aligned} \psi_1 &= -2(\overline{x_1}x_4x_5 + x_1\overline{x_4}x_5 + x_1x_4x_5) \\ &= -2(\overline{x_1}x_4x_5 + x_1x_5) \end{aligned} \qquad (11.3)$$

and from Equation 11.2,

$$E_{2,4} = \psi_1 + h_1 \qquad (11.4)$$

The terms of $E_{2,4}$ are obtained from $E_{1,4}$ (Table 11.2) as follows. Remove the row for block C_1. Since the two terms in ψ_1 (Equation 11.3) involve variables from block C_2, add these two terms to row C_2. The complete $E_{2,4}$ is shown in Table 11.4. This eliminates block C_1. Thus, minimization of $E_{1,4}$ is reduced to minimization of $E_{2,4}$. Removal of vertex C_1 from the partition π_1 gives the new partition $\pi_2 = \{C_2, C_3, C_4\}$ and the corresponding graph G_{π_2} is shown in Figure 11.2(b).

Lemma 11.2: *If graph G_{π_1} is a partial k–tree with elimination sequence $(C_1, C_2, \ldots, C_t)$ then G_{π_2} is also a partial k–tree with elimination sequence $(C_2, C_3, \ldots, C_t)$.*

Proof: Let H denote a k–tree containing G_{π_1}, the graph corresponding to function $E_{1,t}$. Also, let $(C_1, C_2, \ldots, C_t)$ be a k–perfect elimination scheme of H and $N_1 = \{i : C_i$ is connected to $C_1\}$ be the set of neighbors of block C_1 in G_{π_1}. Notice that N_1 induces a complete subgraph in H. Moreover, G_{π_2} is identical to $G_{\pi_1} \setminus C_1$, except possibly for some edges between vertices of N_1 (since ψ_1 depends only on the variables in N_1). Therefore, G_{π_2} is also a subgraph of $H \setminus C_1$, and the claim follows. ∎

Lemma 11.3: *A minimizing point of the function $E_{1,t}$ can be obtained in $O(g^2)$ time.*

Proof: Using Lemma 11.1, we can eliminate block C_1 to generate the new function $E_{2,t}$ in $O(g)$ time. From Lemma 11.2, the elimination sequence for the corresponding graph G_{π_2} is $(C_2, C_3, \ldots, C_t)$. Continuing the elimination process in Lemma 1 for blocks $C_2, C_3, \ldots, C_{t-k}$, successively, we produce two sequences of functions $E_{2,t}, E_{3,t}, \ldots, E_{t-k+1,t}$ and $\psi_1, \psi_2, \ldots, \psi_{t-k}$ where $E_{i,t}, (1 \leq i \leq t - k + 1)$ does not depend on any variable in blocks $C_1, C_2, \ldots, C_i$. From Lemma 1, elimination of a single block involves $O(g)$ work. Since the number of blocks is less than g, the work in eliminating the blocks $\{C_i : 1 \leq i \leq t - k\}$ is bounded by $O(g^2)$. The function $E_{t-k+1,t}$ depends at most on k blocks and the minimum value and a minimizing point of $E_{t-k+1,t}$ can be obtained by examining at most a constant number (2^{kK}) of combinations of values for the variables in the k blocks that E_{t-k+1} depends on. This can be done in $O(g)$ time. Note that the minimum value of the function E_{t-k+1} is also the minimum value of $E_{1,t}$. Given a minimizing point of the function $E_{i+1,t}$, a minimizing point of function $E_{i,t}$ can be obtained in $O(g)$ time using back substitution and, therefore, a minimizing point of $E_{1,t}$ can be obtained in $O(g^2)$ from the minimizing point of $E_{t-k+1,t}$. Hence, a minimizing point of $E_{1,t}$ can be obtained in $O(g^2)$ time. ∎

Example 11.2: Continuing Example 11.1, a minimizing point of $E_{1,4}$ is found by eliminating the remaining blocks. C_2 is eliminated first. The row for C_2 in Table 11.4 gives f_2. The sum of the terms in the remaining rows is h_2. $N_2 = \{C_3, C_4\}$ is the set of neighbors of block C_2 and $J_2 = \{x_1, x_2, x_3\}$. The minimum value of f_2 for each combination of values of variables in J_2 is given in Table 11.5. Therefore, from Equation 11.1,

$$\begin{aligned} \psi_2 &= -2x_2 x_3 (x_1 + \overline{x_1}) \\ &= -2x_2 x_3 \end{aligned} \tag{11.5}$$

and from Equation 11.2,

$$\begin{aligned} E_{3,4} &= \psi_2 + h_2 \\ &= -2x_2 x_3 + 2x_2 x_3 \\ &= 0 \end{aligned}$$

Minimization of $E_{2,4}$ is now reduced to the minimization of $E_{3,4}$ which does not depend on the variables x_4 and x_5. In fact, $E_{3,4} = 0$ and hence, it does not depend on any variable. Clearly, the minimum value of $E_{3,4}$ is 0. Values of variables x_1, x_2 and x_3 are normally determined from $E_{3,4}$. However, since $E_{3,4} = 0$, they can assume any values, say, $x_1 = x_2 = x_3 = 1$ and there exists a test for the output s-a-1 for any values of x_1, x_2 and x_3.

With the elimination phase over, we proceed to find a minimizing point of $E_{2,4}$ from the minimizing point of $E_{3,4}$. From Table 11.5, $f_2(x_1 = x_3 = x_4 = 1) = 2x_4x_5 - 2x_5$ which assumes a minimum value of -2 when $x_4 = 0$ and $x_5 = 1$. Hence, $x_1 = x_2 = x_3 = x_5 = 1, x_4 = 0$ is one minimizing point of $E_{2,4}$. Similarly, we can find a minimizing point of $E_{1,4}$ from that of $E_{2,4}$. From Table 11.3, $f_1(x_1 = 1, x_4 = 0, x_5 = 1) = 2x_6x_7 - 2x_6 + 2x_7$ which assumes a minimum value of -2 when $x_6 = 1$ and $x_7 = 0$. Therefore, $x_1 = x_2 = x_3 = x_5 = x_6 = 1, x_4 = x_7 = 0$ is a minimizing point of $E_{1,4}$ or E_{CKT}. The input $x_1 = x_2 = x_3 = 1, x_4 = 0$ is a test for the primary output s-a-1 fault.

Theorem 11.1: *There exists a polynomial time $(O(g^2))$ algorithm that either detects any primary output fault in a (k, K)–circuit or identifies the fault as redundant.*

Proof: We give a constructive proof. Let C be a (k, K)–circuit with a partition $\pi_1 = \{C_1, C_2, \ldots, C_t\}$ and let $(C_1, C_2, \ldots, C_t)$ be the k–perfect elimination scheme of the graph G_{π_1}. The algorithm is as follows:

1. Construct the energy function E_{CKT} corresponding to the stuck-at primary output fault in circuit C. This can be done in $O(g)$ time (see Chapter 6).

TABLE 11.5: Possible values of f_2.

x_1	x_2	x_3	f_2	min f_2
0	0	0	$6x_5$	0
0	0	1	$2x_5$	0
0	1	0	$2x_5$	0
0	1	1	$-2x_5$	-2
1	0	0	$2x_4x_5 + 6x_5$	0
1	0	1	$2x_4x_5 + 2x_5$	0
1	1	0	$2x_4x_5 + 2x_5$	0
1	1	1	$2x_4x_5 - 2x_5$	-2

2. Find the minimum value of E_{CKT} and the corresponding minimizing point. Using Lemma 11.3, this can be done in $O(g^2)$ time.

3. Check whether minimum value of $E_{CKT} = 0$. If so, the minimizing point corresponds to a test. Otherwise, identify the fault as redundant.

Clearly, the above algorithm runs in $O(g^2)$ time.   ■

11.3.2 Arbitrary Single Fault

We assume that the (k, K)–circuit C only has one primary output. The case of more than one primary output will be discussed later. To detect an arbitrary fault in C, we construct the fault-free and faulty circuit as described in Section 9.1. Since the circuit has just one output, we can avoid the use of exclusive-OR gates. An inverter between the outputs of the fault-free and faulty circuit will ensure that the two circuits produce different outputs (Chapter 6). This inverter is treated as a part of the block that contains the primary output signal. This circuit, as shown in Figure 11.4, is called the *ATG*[2] *circuit*. Note that the inverter is used only for the neural network model which contains bidirectional links.

Lemma 11.4: *The ATG circuit is a $(k, 2K)$-circuit.*

Proof: We will show that the ATG circuit is a $(k, 2K)$-circuit by constructing a partition π such that G_π is a partial k–tree graph. Let $\pi_1 = \{C_1, C_2, \ldots, C_t\}$ be the partition of the fault-free (k, K)–circuit C. We partition the ATG circuit as follows. Observe that for every block C_i in the fault-free circuit lying on a path from the fault site to the primary output, there is a corresponding block C_i' in the faulty circuit. We merge these two blocks into one block M_i. Therefore, $M_i = C_i \cup C_i'$. Since C_i and C_i' each have at most K inputs, the new block M_i will have at most $2K$ inputs. Also, if a block C_j is not on a path from the fault site to a primary output, $M_j = C_j$. Therefore, $\pi = \{M_1, M_2, \ldots, M_t\}$ is a partition of the ATG circuit and G_π is clearly a partial k–tree graph since it is isomorphic to the partial k–tree graph G_{π_1}.   ■

Example 11.3: The ATG circuit for a s-a-0 fault on signal x_6 in Figure 11.1 is shown in Figure 11.4. Only block C_1 is on a path from the fault site to the primary output. The corresponding block in the faulty circuit is C_1'.

---

[2]Automatic Test Generation.

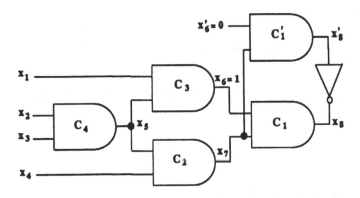

FIGURE 11.4: ATG circuit for x_6 s-a-0 in the example of Figure 11.1.

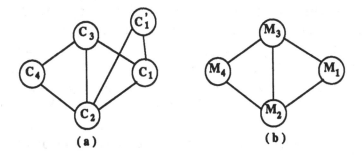

FIGURE 11.5: Graphs $G_{\pi'}$ (a) and G_{π} (b) for the single fault ATG circuit.

The graph $G_{\pi'}$ corresponding to the partition $\pi' = \{C_1, C_2, C_3, C_4, C_1'\}$ is shown in Figure 11.5(a). Merging blocks C_1 and C_1' causes the edges (C_1, C_1') and (C_2, C_1') in graph $G_{\pi'}$ to disappear. Also, $M_1 = C_1 \cup C_1'$. Since none of the other blocks in the fault-free circuit are on a potential fault propagation path, $M_2 = C_2$, $M_3 = C_3$ and $M_4 = C_4$. The graph G_{π} corresponding to the partition $\pi = \{M_1, M_2, M_3, M_4\}$ is shown in Figure 11.5(b). Clearly, G_{π} is isomorphic to the partial k-tree graph G_{π_1} in Figure 11.2(a).

Theorem 11.2: *There exists a polynomial time $(O(g^2))$ algorithm that either detects any arbitrary given fault in a (k, K)-circuit or identifies the fault as redundant.*

Proof: We give a constructive proof. Let signal x_i be s-a-0(1). The following algorithm either detects the fault or identifies it as redundant:

    1. Construct the energy function E_{ATG} corresponding to the ATG circuit for the fault. Simplify E_{ATG} by substituting $x_i = 1(0)$ and

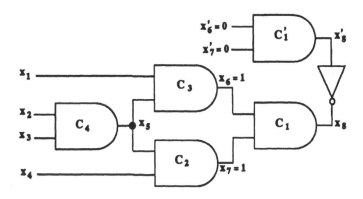

FIGURE 11.6: ATG circuit for the multiple fault x_6 s-a-0, x_7 s-a-0.

$x_i' = 0(1)$ where x_i and x_i' are the signals in the fault-free and faulty circuits, respectively, corresponding to the fault site. Also, from Lemma 11.4, the ATG circuit is a partial k–tree with $(M_1, M_2, \ldots, M_t)$ as a perfect elimination scheme.

2. Find a minimizing point of E_{ATG}. From Lemma 11.3, such a point can be obtained in $O(g^2)$ time.

3. Check whether at the minimizing point $E_{ATG} = 0$. If so, the minimizing point corresponds to a test. Otherwise, identify the fault as redundant.

Clearly, the above algorithm runs in $O(g^2)$ time.   ∎

Multi-output circuits: For (k, K)–circuits with more than one primary output, we repeat the above procedure for each primary output that is reachable from the fault site. Since the number of primary outputs of the circuit is bounded by g (the number of signals in the circuit), the procedure may have to be repeated only a polynomial number of times. An alternative method would be to construct the ATG circuit for the multi-output circuit as described in Chapter 6, find an elimination scheme, and use the procedure once to generate a test.

11.3.3 Multiple Faults

We again assume that the (k, K)–circuit C only has one primary output. The case of more than one primary outputs will be discussed later. To detect a multiple fault in C, we construct the fault-free and faulty circuit in a similar fashion as described in Sections 9.1 and 11.3.2 except that we duplicate all blocks that are on the path from any of the multiple fault sites. For

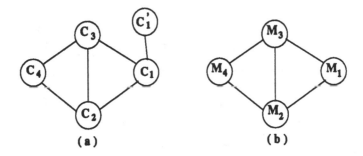

FIGURE 11.7: Graphs $G_{\pi'}$ (a) and G_{π} (b) for the multiple fault ATG circuit.

example, the ATG circuit for the circuit of Figure 11.1 with a multiple fault on signals x_6 s-a-0 and x_7 s-a-0 is shown in Figure 11.6. Block C_1 is duplicated and $\pi' = \{C_1, C_2, C_3, C_4, C_1'\}$ is a partition of the ATG circuit. The corresponding graph $G_{\pi'}$ is shown in Figure 11.7(a). Merging the blocks on the fault propagation paths gives the partition $\pi = \{M_1, M_2, M_3, M_4\}$. The corresponding graph is shown in Figure 11.7(b). From Lemma 11.4, it follows that the ATG circuit is a $(k, 2K)$-circuit.

Theorem 11.3: *There exists a polynomial time $(O(g^2))$ algorithm that either detects any multiple fault in a (k, K)–circuit or identifies the fault as redundant.*

Proof: The following algorithm either detects a multiple fault or identifies it as redundant:

1. Construct the energy function E_{ATG} for the ATG circuit of the multiple fault. Simplify E_{ATG} by substituting appropriate values for the variables in the fault-free and faulty circuits corresponding to the multiple fault sites. Also, from Lemma 11.4, the ATG circuit is a partial k–tree with $(M_1, M_2, \ldots, M_t)$ as the perfect elimination scheme.

2. Find a minimizing point of E_{ATG}. From Lemma 11.3, such a point can be obtained in $O(g^2)$ time.

3. Check whether the minimum value of $E_{ATG} = 0$. If so, the minimizing point corresponds to a test. Otherwise, identify the fault as redundant.

Clearly, the above algorithm runs in $O(g^2)$ time.   ∎

Multi-output circuits: For (k, K)–circuits with more than one primary output, we repeat the above procedure for each primary output that is reachable from the fault site. Since the number of primary outputs of the circuit is bounded by g (the number of signals in the circuit), the number of times the above procedure may be repeated is bounded by a polynomial in g ($O(g)$). Again, an alternative method would be to construct the ATG circuit for the multi-output circuit as described in Chapter 6, find an elimination scheme, and use the procedure once to generate a test.

11.4 Summary

We have identified a class of combinational circuits in which the single and multiple fault detection problem can be solved in polynomial time. A novel test generation method is presented. This is a significant result in test generation and in the identification of redundancies. The latter has important applications in the synthesis of logic circuits. Future research should extend the fault detection algorithm for (k, K)–circuits to general combinational circuits. Also, techniques to synthesize combinational functions as easier-to-test (k, K)–circuits should be investigated.

References

[1] V. D. Agrawal and S. C. Seth. *Test Generation for VLSI Chips*. IEEE Computer Society Press, Los Alamitos, CA, 1988.

[2] A. V. Aho, J. E. Hopcroft, and J. D. Ullman. *The Design and Analysis of Computer Algorithms*. Addison-Wesley Publishing Company, Reading, MA, 1974.

[3] S. T. Chakradhar, V. D. Agrawal, and M. L. Bushnell. Polynomial Time Solvable Fault Detection Problems. In *Proceedings of the 20th IEEE International Symposium on Fault Tolerant Computing*, pages 56–63, June 1990.

[4] H. Fujiwara. Computational Complexity of Controllability/Observability Problems for Combinational Circuits. *IEEE Transactions on Computers*, C-39(6):762–767, June 1990.

[5] O. H. Ibarra and S. K. Sahni. Polynomially Complete Fault Detection Problems. *IEEE Transactions on Computers*, C-24(3):242–249, March 1975.

[6] D. J. Rose. On Simple Characterizations of k-trees. *Discrete Mathematics*, 7(3,4):317–322, February 1974.

Chapter 12

SPECIAL CASES OF HARD PROBLEMS

"Under the worst-case complexity measure, most design automation problems are intractable. This conclusion remains true even if we are interested only in obtaining solutions with values guaranteed to be 'close' to the value of the optimal solutions. The most promising approaches to certifying the value of algorithms for these intractable problems appear to be probabilistic analysis and experimentation. Another avenue of research that may prove fruitful is the design of highly parallel algorithms (and associated hardware) for some of the computationally more difficult problems."

– S. Sahni in *Handbook of Advanced Semiconductor Technology and Computer Systems*, Van Nostrand Reinhold (1988)

It is unlikely that a polynomial time algorithm exists for finding exact solutions of all instances of an NP-complete problem [8]. Heuristics are generally used to find reasonably good solutions. However, there may be several instances of the problem for which exact solutions can be found quickly (i.e., in polynomial time). Identification of such instances is useful for two reasons:

141

1. They can be solved very quickly.

2. Solution methods for these instances can provide better insights into the general problem.

In this chapter, we identify polynomial-time solvable special instances of an NP-complete problem [5]. We have included sufficient background material to make this chapter self-contained. However, a familiarity with our neural network model of logic circuits given in Chapter 5 will be helpful.

12.1 Problem Statement

We will consider minimization of a *pseudo-Boolean quadratic function* (i.e., quadratic 0-1 programming) as a representative NP-complete problem. A pseudo-Boolean quadratic function $f : \{0, 1\}^n \to \Re$ is defined as $f(\mathbf{x}) = \mathbf{x}^T\mathbf{Q}\mathbf{x} + \mathbf{c}^T\mathbf{x}$, where $\mathbf{Q} = [Q_{ij}]$ is an $n \times n$ symmetric matrix of constants with null elements in the diagonal, $\mathbf{c}$ is a column vector of n constants, $\mathbf{x}$ is a column vector of n binary (0 or 1) variables and $\mathbf{x}^T$ is the transpose of $\mathbf{x}$. There is no loss of generality due to the null diagonal assumption because $x_i^2 = x_i$ for $1 \leq i \leq n$.

We associate a graph $G_f(\mathcal{V}, \mathcal{E})$ ($\mathcal{V}$ is the vertex set and $\mathcal{E}$ the edge set of the graph) with f as follows:

$$\mathcal{V} = \{x_1, x_2, \ldots, x_n\}$$
$$(x_i, x_j) \in \mathcal{E} \Leftrightarrow Q_{ij} \neq 0 \ (i \in \mathcal{V}, j \in \mathcal{V})$$

If $e \in \mathcal{E}$ is an edge with extremities x_i and x_j we write (x_i, x_j) to denote the edge e. Also, edge (x_i, x_j) is said to be *incident* on vertices x_i and x_j. Many combinatorial optimization problems like the VLSI layout problem, the traveling salesperson problem, the max-cut problem and others can be solved via quadratic 0-1 programming [10].

Finding the minimum of f is an NP-complete problem [8]. In this chapter, we use the phrases *minimum of f* and *global minimum of f*, interchangeably. For the general case, a branch-and-bound algorithm has been proposed by Carter [4] and polynomial algorithms for computing lower bounds of f have been given by Hammer *et al* [9]. However, this problem is solvable in polynomial time when all the elements of Q are nonpositive [12]. Barahona [2] showed that if the graph G_f is series-parallel (planar graphs that are not reducible to a fully connected graph of four vertices are series-parallel graphs), then the minimum of f can be obtained in linear time. This result includes, as a special case, the maximum stable set problem in series-parallel graphs that was solved by Boulala and Uhry [3].

In this chapter, we present a new class of quadratic 0-1 programming cases solvable in $O(n)$ time where n is the number of variables in the pseudo-Boolean quadratic function. We show that if the graph G_f is transformable into a combinational circuit of *inverters* and 2-input *AND*, *OR*; *NAND* and *NOR logic gates*, then the minimum of f can be obtained in linear time. A novel modeling technique (described in Chapter 5) is used to transform G_f into a logic circuit. The minimum of f is determined by identifying the *primary input* signals of the circuit and by performing *logic simulation* for an arbitrary set of 0-1 values for these input signals.

A basic knowledge of logic circuits and neural networks is required to understand the techniques presented in the sequel. For a review of logic circuits and neural networks, refer to Chapters 2 and 3, respectively. In Section 12.2, we review logic simulation. Neural modeling as discussed in Chapter 5 is crucial to transforming G_f into a logic circuit. To make this chapter self-contained, we review the relevant details of the modeling technique in Section 12.3. All instances of the logic simulation problem can be formulated as quadratic 0-1 programs but only a subset of quadratic 0-1 programming cases can be formulated as logic simulation problems. The formulation of the logic simulation problem as a quadratic program gives valuable insights for developing a procedure to isolate the subset of quadratic programs that can be formulated as logic simulation problems. Section 12.4 formulates the logic simulation problem as a quadratic 0-1 programming problem and Section 12.5 presents a linear time algorithm to verify whether or not a given quadratic 0-1 program can be formulated as a logic simulation problem. If a formulation exists, the algorithm constructs the logic circuit C corresponding to G_f. Logic simulation of C for an arbitrary set of primary input values yields a minimizing point of the pseudo-Boolean function.

In Section 12.6 we investigate the complexity of solving restricted versions of the quadratic 0-1 programming problem and prove that the problem of finding the minimum of f, even in the restricted case when all elements of Q are positive, is NP-Complete. The proof establishes an interesting connection between quadratic 0-1 programming and test generation for digital circuits.

12.2 Logic Simulation

Consider a combinational circuit C consisting of signals $x_1, x_2, \ldots, x_n$. Let $H_C(\mathcal{V}_C, \mathcal{E}_C)$, where $\mathcal{V}_C$ is the vertex set and $\mathcal{E}_C$ the edge set, be the *circuit graph* associated with C. $\mathcal{V}_C = \{x_1, x_2, \ldots, x_n\}$ and there is an arc $(x_i, x_j) \in \mathcal{E}_C$ ($1 \leq i \leq n$ and $1 \leq j \leq n$) if there is a gate x_j with x_i

as an input signal. If H_C has no directed *cycles* [1], C is a combinational circuit. Otherwise, C is a sequential circuit. For example, it can easily be verified that the circuit graph corresponding to the circuit in Figure 12.1 has no directed cycles and, therefore, it is a combinational circuit. In the sequel, we will only deal with combinational circuits.

Without loss of generality, let x_n be the primary output of C. The signals in the circuit can always be relabeled to achieve this. We define C^n as the circuit obtained by *deleting* gate x_n from C. More formally, if

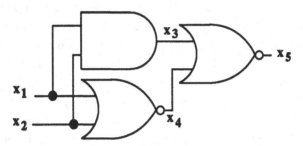

FIGURE 12.1: An example logic circuit.

$H_C(\mathcal{V}_C, \mathcal{E}_C)$ is the circuit graph of C, the circuit graph $H_{C^n}(\mathcal{V}_{C^n}, \mathcal{E}_{C^n})$ of circuit C^n is defined as follows: $\mathcal{V}_{C^n} = \{x_1, x_2, \ldots, x_{n-1}\}$ and there is an arc $(x_i, x_j) \in \mathcal{E}_{C^n}$ if $(x_i, x_j) \in \mathcal{E}_C$ ($1 \leq i \leq n-1$ and $1 \leq j \leq n-1$). For example, x_5 is the primary output of the circuit in Figure 12.1 and the circuit obtained by deleting gate x_5 is shown in Figure 12.2.

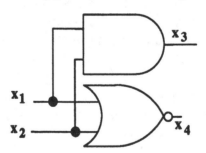

FIGURE 12.2: Logic circuit obtained by deleting gate x_5 from Figure 12.1.

Logic simulation is the process of computing the values of the primary output signals, given a set of values for the primary inputs. For combinational circuits, logic simulation for an input value set can be performed in time complexity that is linear in the number of signals in the logic circuit. The first step in logic simulation is *levelizing*. Let $P_i = \{x_j : \text{arc } (x_j, x_i) \in \mathcal{E}_C, 1 \leq j \leq n\}$ as the set of *predecessors* of signal x_i in graph H_C. Therefore, P_i is the set of input signals to gate x_i. We define the *level* of a signal z_i in combinational circuit C as follows:

- level$(z_i) = 0$ for all primary input signals.

- level$(z_i) = \max\{$level$(z_j) : z_j \in P_i\} + 1$ for all other signals.

Example 12.1: In Figure 12.1, level$(z_1) = 0$ and level$(z_2) = 0$ since z_1 and z_2 are primary input signals. Also, $P_3 = P_4 = \{z_1, z_2\}$ and $P_5 = \{z_3, z_4\}$. Therefore, level$(z_3) = \max\{$level$(z_1),$ level$(z_2)\} + 1 = 1$. Similarly, level$(z_4) = 1$ and level$(z_5) = 2$.

The level of every signal in the circuit can be computed in time complexity that is linear in the number of signals in the circuit [7]. Given a set of values for the primary inputs of C, logic simulation is performed as follows:

1. *Levelize* the circuit. Note that a signal is assigned a higher level number than all its predecessors. As stated earlier, the complexity of levelizing is $O(n)$.

2. Compute the values of all unknown signals in ascending order of levels. The value of a signal z_i can easily be determined from the signal values of its predecessors. For example, if z_i is the output signal of an AND gate and P_i is the set of input signals to this gate, then z_i is the conjunction of values of signals in P_i. Each signal value can be uniquely determined because when the value of a signal z_i is computed the signal values of all its predecessors have already been computed. Since each gate is visited only once in this process, the complexity of this step is $O(n)$.

Clearly, the work involved in logic simulation is $O(n)$, where n is the number of signals in the circuit.

Example 12.2: Given that the primary input signals z_1 and z_2 are 0 in the circuit in Figure 12.1, we will perform logic simulation to compute the values of all other signals. We compute the levels of all signals as shown in Example 12.1 and process the signals in the ascending order of levels: z_3, z_4, z_5 (note that we could have also processed the signals in the order z_4, z_3, z_5 since z_4 and z_3 have the same level number). Since z_3 is the output of an AND gate, $z_3 = z_1 \wedge z_2 = 0 \wedge 0 = 0$. Similarly, $z_4 = \overline{z_1 \vee z_2} = \overline{0 \vee 0} = 1$ and $z_5 = \overline{z_3 \vee z_4} = \overline{0 \vee 1} = 0$.

Consider the AND gate with z_1 and z_2 as its inputs and z_3 as the output. A set of values of z_1, z_2 and z_3 is *consistent* with the function of the AND gate if $x_3 = x_1 \wedge x_2$. Otherwise, the set of values is *inconsistent*. For example, $x_1 = x_2 = x_3 = 0$ is consistent with the function of the AND gate but $x_1 = x_2 = 0, x_3 = 1$ is inconsistent. An assignment of signal values that is consistent with the function of all gates in the circuit is called a *consistent labeling* of the circuit. For example, signal values determined in Example 12.2 constitute a consistent labeling of the circuit in Figure 12.1 since $(x_1 = x_2 = x_3 = 0)$ is consistent with the function of the AND gate x_3, $(x_1 = x_2 = 0, x_4 = 1)$ is consistent with the function of the NOR gate x_4 and $(x_3 = x_5 = 0, x_4 = 1)$ is consistent with the the function of the NOR gate x_5. Similarly, performing logic simulation with $x_1 = x_2 = 1$ results in the consistent labeling, $x_1 = x_2 = x_3 = 1, x_4 = x_5 = 0$. In general, a circuit with p primary inputs has 2^p consistent labelings since performing logic simulation with every set of primary input signal values results in a consistent labeling of the circuit.

12.3   Logic Circuit Modeling

For an understanding of the the present discussion, we require some knowledge of neural networks and their use in the modeling of combinational circuits. In this section, therefore, we summarize some essentials from Chapters 4 and 5.

A neural network is a network of computing elements (*neurons*) whose function is characterized by a pseudo-Boolean quadratic function, called the *energy* function. The neural network is uniquely specified by the weights on the connections between the neurons and by the thresholds of the neurons. For an n-neuron network, the weights are represented as an $n \times n$ matrix $\mathbf{T}$ and the thresholds, as an n-component vector $\mathbf{I}$. The energy function is of the form [11]:

$$E = -[1/2 \sum_{i=1}^{n} \sum_{j=1}^{n} T_{ij} x_i x_j] - [\sum_{i=1}^{n} I_i x_i] + K \qquad (12.1)$$

where $T_{ij} \in \Re$ is the weight associated with the connection from neuron j to i, x_i is the activation value of neuron i, $I_i \in \Re$ is the threshold of neuron i and K is a constant. We use Hopfield neural networks [11] in which the connections are bidirectional, i.e., if there is a link from neuron i to neuron j, then there exists a link from j to i. Furthermore, the link weights are symmetric (i.e., $T_{ij} = T_{ji}$, $1 \leq i \leq n$, $1 \leq j \leq n$) and $T_{ii} = 0$. We will occasionally refer to neuron i as x_i when there is no confusion between the name of a neuron and its activation value.

$G_E(\mathcal{V}, \mathcal{E})$, the *neural network graph* associated with the energy function E, is defined as follows:

$$\mathcal{V} = \{x_1, x_2, \ldots, x_n\}$$
$$(x_i, x_j) \in \mathcal{E} \Leftrightarrow T_{ij} \neq 0 \quad (i \in \mathcal{V}, j \in \mathcal{V})$$

Furthermore, we associate a *threshold* $I_i \in \Re$ with the vertex $x_i \in \mathcal{V}$ and a *weight* $T_{ij} \in \Re$ with the edge $(x_i, x_j) \in \mathcal{E}$.

12.3.1 Model for a Boolean Gate

Figure 12.3 shows an AND gate and the corresponding energy function is given by:

$$E_{AND}(x_3, x_1, x_2) = -[4x_1x_3 + 4x_2x_3 - 2x_1x_2] - [-6x_3] \tag{12.2}$$

which is derived from Table 5.1, with $A = B = 2$. Variables x_1, x_2 and x_3 can assume only binary values. All operations are arithmetic and not Boolean. It is easily verified that only those values of x_1, x_2 and x_3 that are consistent with the function of the AND gate will satisfy $E_{AND} = 0$. Furthermore, $E_{AND} > 0$ for all other combinations of x_1, x_2 and x_3. Note that E_{AND} is a pseudo-Boolean quadratic function that assumes a minimum value of 0 only at values of x_1, x_2 and x_3 consistent with the function of the AND gate. The neural network graph G_{AND} corresponding to Equation 12.2 is shown in Figure 12.3. Each signal is represented by

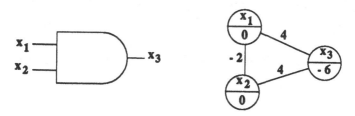

FIGURE 12.3: AND gate and its neural network graph.

a neuron having a threshold. The neurons for signals x_1 and x_2 have 0 thresholds and the neuron for signal x_3 has a threshold of -6. All links are bidirectional and their labels show their weights.

Similar energy functions and the corresponding neural network graphs are derived in Chapter 5 for other logic gates. As shown there, it is not important that E_{AND} or the energy function for other gates have a minimum value of 0. In fact, by the addition of a constant, the energy function can be made to have an arbitrary value as its global minimum. Without loss of generality, in the sequel we assume that the energy function for a gate assumes a minimum value of 0. Neural networks for 2-input AND, OR, NAND, NOR, XOR and XNOR gates and the inverter constitute the *basis set* and any gate with more than two inputs can be constructed from this basis set. Table 5.1 shows **T** and **I** for the neural networks in the basis set.

Neurons corresponding to the primary inputs (outputs) of the circuit are called *primary input (output) neurons*. Neurons corresponding to the input (output) signals of a gate are called *input (output) neurons*. For example, x_1 and x_2 are the input neurons and x_3 is the output neuron in the AND gate neural network shown in Figure 12.3. The energy function for the AND gate, as given by Equation 12.2, can be derived from Table 5.1 by setting $A = B = 2$. Energy functions for the XOR and XNOR gates are given in Table 5.3. From Table 5.1, clearly, the energy function for a gate in the basis set has at most seven terms and has a minimum value 0 at all consistent states of the gate. Let G_{AND}, G_{OR}, G_{NAND}, G_{NOR} and G_{NOT} be the neural network graphs corresponding to the energy functions of the AND, OR, NAND, NOR and NOT gates, respectively.

12.3.2 Circuit Modeling

Consider the three-gate circuit C shown in Figure 12.1. The AND gate energy function is given by Equation 12.2. From Table 5.1, with $A = B = 2$, the energy functions for the two NOR gates are:

$$E_{NOR}(x_4, x_1, x_2) \;=\; 4x_1x_4 + 4x_2x_4 + 2x_1x_2 - 2(x_1 + x_2 + x_4) + 2$$
$$E_{NOR}(x_5, x_3, x_4) \;=\; 4x_3x_5 + 4x_4x_5 + 2x_3x_4 - 2(x_3 + x_4 + x_5) + 2$$

The energy function for the entire circuit C is obtained by summing the individual gate energy functions. Thus,

$$\begin{aligned} E_C \;=\;\; & E_{AND}(x_3, x_1, x_2) + E_{NOR}(x_4, x_1, x_2) \\ & + E_{NOR}(x_5, x_3, x_4) \end{aligned} \qquad (12.3)$$

This simple procedure can model circuits with any connectivity and any types of gates (see Chapter 5). Since the models are defined only for two-input gates, gates with larger fan-in are represented as combinations of two-input gates. Individual gate energy functions only assume non-negative values and, therefore,

- E_C is non-negative.

- $E_C = 0$ only when the individual energy functions separately become 0.

- Any solution of $E_C = 0$ is a consistent labeling for the entire circuit.

Since a logic circuit with p primary inputs has 2^p consistent labelings, the corresponding energy function has 2^p minimizing points. Also, if g is the number of gates in the logic circuit, the energy function for the circuit has at most $7 \times g$ terms since, from Table 5.1, the energy function for each gate in the logic circuit has at most seven terms. Since $g \leq n$ (n being the number of signals in the circuit), the number of terms in the energy function for the circuit is $O(n)$.

12.4 Simulation as a Quadratic 0-1 Program

The logic simulation problem in an arbitrary combinational logic circuit C can be reduced to a quadratic 0-1 optimization problem by transforming the logic circuit into an energy function E_C which assumes a minimum value of 0 only at neuron states consistent with the function of all gates in the logic circuit. Logic simulation will require a solution of $E_C = 0$ when the primary input signals assume given values.

Example 12.3: The energy function for the circuit in Figure 12.1 is given by Equation 12.3. The logic circuit has two primary inputs and, therefore, only four consistent labelings which are shown in Table 12.1. The corresponding values of E_C (Equation 12.3) are shown in the last column. It can easily be verified that E_C assumes the minimum value 0 only at these four consistent labelings of the circuit.

TABLE 12.1: Consistent states of circuit in Figure 12.1.

x_1	x_2	x_3	x_4	x_5	E_C
0	0	0	1	0	0
0	1	0	0	1	0
1	0	0	0	1	0
1	1	1	0	0	0

If a pseudo-Boolean quadratic function f can be transformed into a logic circuit, then the minimum of f can be obtained through logic simulation. In the next section, we present a linear time algorithm to transform f into a combinational circuit.

12.5 Quadratic 0-1 Program as Simulation

As mentioned earlier, only a subset of quadratic 0-1 programs can be formulated as instances of the logic simulation problem. We now present a linear time algorithm that determines whether or not a given quadratic 0-1 program can be thus formulated. An example is provided to illustrate the mechanics of the algorithm.

12.5.1 A Linear Time Algorithm

We assume that the function f has the same form as Equation 12.1. Otherwise, f can be rewritten in the same form as Equation 12.1.

Example 12.4: Consider the following quadratic function:

$$f = -5x_1x_3 + 7x_1x_4 + 7x_1x_2 - 5x_2x_3 + 7x_2x_4 + 3x_3x_5 + 3x_4x_5$$
$$+2x_3x_4 - 4x_1 - 4x_2 + 5x_3 - 6x_4 - 2x_5 + 6$$

It can be rewritten as

$$f = -[5x_1x_3 - 7x_1x_4 - 7x_1x_2 + 5x_2x_3 - 7x_2x_4 - 3x_3x_5 - 3x_4x_5$$
$$-2x_3x_4] - [4x_1 + 4x_2 - 5x_3 + 6x_4 + 2x_5] + 6 \qquad (12.4)$$

The corresponding neural network graph G_f, as defined in Section 12.3, is shown in Figure 12.5. A vertex is indicated by a circle. The vertex label is indicated in the upper half and the vertex-weight is indicated in the lower half. For example, vertex x_3 has a vertex-weight of -5 since $I_3 = -5$. The graph has five vertices since f is a function of five variables $x_1 \ldots x_5$. There is an edge between two vertices x_i and x_j if function f has a term $T_{ij}x_ix_j$ and the edge-weight T_{ij} is indicated as a label on the edge. For example, edge (x_3, x_4) has an edge-weight of -2 since $T_{34} = -2$.

The algorithm attempts to construct a combinational logic circuit from G_f. Each vertex in G_f corresponds to a signal in the logic circuit. The algorithm first deletes vertices and edges of G_f that correspond to logic gates associated with the primary outputs of the logic circuit. Then it recursively constructs the remaining logic gates in the circuit by deleting appropriate edges and vertices in G_f and tracing the signals in the logic circuit from primary outputs to the primary inputs.

We give some graph-theoretic definitions that are used in this section. Let $G(\mathcal{V}, \mathcal{E})$ be a graph with vertex set $\mathcal{V}$ and edge set $\mathcal{E}$. If $\mathcal{W} \subseteq \mathcal{V}$, the subgraph induced by $\mathcal{W}$ is the graph $H(\mathcal{W}, \mathcal{F})$ such that $(i, j) \in \mathcal{F}$ if and only if $(i, j) \in \mathcal{E}$ and $\{i, j\} \subseteq \mathcal{W}$. For $v \in \mathcal{V}$, $G \setminus v$ will denote the subgraph induced by the set $\mathcal{V} \setminus v$ ($\setminus$ denotes the set minus operation).

We make the following observations regarding a neural network graph of any combinational circuit. Let C be a circuit with signals $x_1 \ldots x_n$ and let x_n be a primary output. Let E_C be the energy function of circuit C and G_C be the associated neural network graph.

Lemma 12.1: *The neural network graph G_C has at least one vertex v of degree one or two. Moreover, if v is of degree two, both edges incident on v will have equal weights.*

Proof: Since x_n is the primary output of circuit C, it is a primary output neuron in G_C. Furthermore, x_n is not an input signal to any gate in the circuit and gate x_n has at most two inputs. Therefore, x_n has degree one (i.e., it is a signal associated with an inverter) or two in G_C. If the degree of x_n is two, then the weights on the edges incident on x_n are equal because the output neuron of a neural network graph of every logic gate in the basis set has two equally-weighted edges (Table 5.1). ∎

Using Lemma 12.1, all the primary outputs of circuit C can be easily identified from the graph G_C. A vertex v in G_C corresponds to a primary output of circuit C if

- the degree of vertex v is one, or

- the degree of vertex v is two and both edges incident on v have equal weights.

Lemma 12.2: *Let x_n be a primary output neuron in G_C. The threshold of x_n and the weight on an edge incident to x_n uniquely determine*

- *the gate-type (i.e., AND, OR, NAND, NOR or NOT gate) of gate x_n, and*

- *the constants A, B and J (Table 5.1) associated with the neural network graph of gate x_n.*

Proof: Let I_n be the threshold of x_n and w be the weight on an edge incident to x_n. The degree of x_n is at most two (Lemma 12.1). If the degree of x_n is one, then the logic gate must be an inverter since only G_{NOT} has a primary output neuron of degree one. Therefore, from Table 5.1, $-2J = w$ and $J = I_n$, i.e., $J = \frac{-w}{2} = I_n$. If the degree of x_n is two, then both edges incident to x_n have equal edge-weights (Lemma 12.1). We have two cases:

CASE 1: $w > 0$.

From Table 5.1, it can easily be verified that only in G_{AND} and G_{OR} do the edges incident on the primary output neuron have positive edge-weights. Therefore, gate x_n could be an AND or OR gate. Again, from Table 5.1, if $w < -I_n$, then the gate is AND type with $A + B = w$ and $-(2A + B) = I_n$, i.e., $A = -(w + I_n)$, $B = 2w + I_n$. If $w > -I_n$, then the gate is OR type with $A + B = w$ and $-B = I_n$, i.e., $A = w + I_n$, $B = -I_n$.

CASE 2: $w < 0$.

The gate is NAND or NOR type. From Table 5.1, if $-w < I_n$, then the gate is NAND type with $-A - B = w$ and $2A + B = I_n$, i.e., $A = w + I_n$, $B = -2w - I_n$. If $-w > I_n$, then the gate is NOR type with $-A - B = w$ and $B = I_n$, i.e., $A = -w - I_n$. ∎

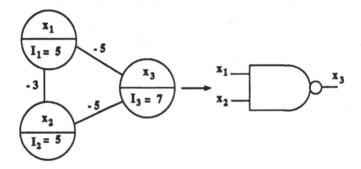

FIGURE 12.4: A neural network graph that represents a logic gate.

Example 12.5: As an illustration, consider the graph shown in Figure 12.4. Vertex x_3 is the primary output neuron since it has two equally-weighted edges incident on it. The threshold of x_3 is 7, i.e., $I_3 = 7$, and the weight on an edge incident to x_3 is -5, i.e., $w = -5$. Since, $w < 0$, gate x_3 could possibly be a NAND or NOR gate. However, $-w < I_3$ and therefore, gate x_3 is a NAND gate with $A = w + I_3 = 2$ and $B = -2w - I_3 = 3$.

Let C^n be the circuit obtained by deleting gate x_n from circuit C. Let E_{C^n} and G_{C^n} be the energy function and neural network graph, respectively, of circuit C^n.

Lemma 12.3: G_{C^n} *can be constructed from* G_C *in* $O(1)$ *time.*

Proof: Using Lemma 12.2, we can determine the gate type of gate x_n and the constants A, B and J. Let E_{x_n} be the energy function of gate x_n.

The proof is based on the following observation: Since C^n is obtained by deleting gate x_n from circuit C, $E_{C^n} = E_C - E_{x_n}$. Therefore, we can construct G_{C^n} from G_C by deleting vertex x_n and the edges incident on vertex x_n, and by appropriately modifying:

1. the thresholds of neighbors of x_n, and

2. the edge-weights on the edges between the neighbors of x_n, so that G_{C^n} is the graph of the energy function E_{C^n}.

The degree of x_n is at most two (Lemma 12.1). If the degree of x_n is one, let x_q be the neighboring vertex in graph G_C and I_q be the threshold of vertex x_q. Since circuit C^n has all signals in circuit C except signal x_n, $G_{C^n} = G_C \setminus x_n$. Furthermore, removal of gate x_n changes the threshold of vertex x_q to $I_q - J$. Clearly, these modifications to G_C can be made in $O(1)$ time.

If x_n is of degree two, let x_q and x_r be the two vertices connected to x_n, and let I_q and I_r be their respective thresholds. Let T_{qr} be the edge-weight on edge (x_q, x_r) in graph G_C. $G_{C^n} = G_C \setminus x_n$ and the thresholds of x_q and x_r, and the edge-weight T_{qr}, are modified depending on the gate-type of gate x_n. If gate x_n is an AND gate, I_q and I_r are unchanged and $T_{qr} - (-B) = T_{qr} + B$ is the new edge-weight on edge (x_q, x_r). This is because, from Table 5.1, the AND gate x_n contributes $-B$ to the edge-weight of edge (x_q, x_r) in G_C but it does not contribute to the thresholds of vertex q and r. Similarly, if gate x_n is

- OR type, I_q, I_r and T_{qr} are modified to $I_q + A$, $I_r + A$ and $T_{qr} + B$, respectively.

- NAND type, I_q, I_r and T_{qr} are modified to $I_q - A - B$, $I_r - A - B$ and $T_{qr} + B$, respectively.

- NOR type, I_q, I_r and T_{qr} are modified to $I_q - B$, $I_r - B$ and $T_{qr} + B$, respectively.

∎

In the sequel, the process of constructing G_{C^n} from G_C will be referred to as the *deletion* of gate x_n from G_C.

Given a function f and its graph G_f, we construct a combinational circuit as follows: we assume that G_f represents some combinational circuit and use Lemma 12.3 to recursively delete primary output neurons from G_f and the successive graphs obtained by deletion.

Theorem 12.1: *Let $f(\mathbf{x}) = \mathbf{x}^T Q \mathbf{x} + \mathbf{c}^T \mathbf{x}$ be a pseudo-Boolean quadratic function of n variables. Whether or not G_f is transformable to a combinational circuit can be determined in $O(n)$ time.*

Proof: We rewrite f so that it is in the same form as Equation 12.1 and let G_f be the graph associated with f. We provide a constructive proof. The construction of the combinational circuit can be performed in the following steps:

1. For every vertex v in G_f, compute its degree d_v and put in a list L all vertices having either (i) $d_v \leq 1$ or (ii) $d_v = 2$ and both edges incident on v having the same weight. If there exists a combinational circuit corresponding to G_f, then L is not empty (Lemma 12.1). Clearly, if there exists a combinational circuit for f, all vertices in L would be the primary outputs of the circuit.

2. Choose a vertex v in L.

   CASE 1: $d_v = 2$.

   Let y and z be the vertices adjacent to v. Using Lemma 12.3, delete gate v from G_f. This appropriately modifies the thresholds of y and z and the edge weight on edge (y, z), depending on the gate type of v. If edge weight on edge (y, z) is 0, then set $d_y \leftarrow d_y - 2, d_z \leftarrow d_z - 2$. Otherwise, set $d_y \leftarrow d_y - 1, d_z \leftarrow d_z - 1$.

   Include y in L if either (i) $d_y \leq 1$ or (ii) $d_y = 2$ and both edges incident on y have the same weight. Similarly, include z in L if either (i) $d_z \leq 1$ or (ii) $d_z = 2$ and both edges incident on z have the same weight.

   CASE 2: $d_v = 1$.

   Let y be the vertex adjacent to v. Using Lemma 12.3, delete gate v from G_f. This modifies the threshold of vertex y.

   Set $d_y \leftarrow d_y - 1$, and if $d_y \leq 2$ we include y in L.

   CASE 3: $d_v = 0$.

   If the threshold of v is not 0, then G_f does not correspond to a combinational circuit and we stop. Otherwise mark v as a primary input of the combinational circuit.

3. Set $G_f \leftarrow G_f \setminus v$ and $L \leftarrow L \setminus v$. Repeat Step 2 until L is empty.

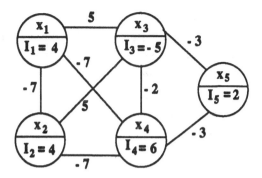

FIGURE 12.5: The graph G_f of function f in Example 12.4.

The work needed by this algorithm is bounded by the number of edges in the graph G_f. If G_f corresponds to a combinational circuit, then G_f has $O(n)$ edges where n is the number of vertices in G_f. If G_f does not correspond to a combinational circuit, the algorithm terminates, either in Step 1 or in Case 3 of Step 2, after examining at most $O(n)$ edges. ∎

Example 12.6: As an illustration of Theorem 12.1, consider the quadratic 0-1 function f given by Equation 12.4. Figure 12.5 shows the graph G_f associated with f. Note that G_f is not a series-parallel graph. The transformation of G_f into a combinational circuit is performed as follows:

1. Construct the list L. It consists of a single element, vertex x_5.

2. Remove vertex x_5 from L, i.e., $L \leftarrow L \setminus x_5$.

3. Delete gate x_5 from G_f using Lemma 12.3. From Lemma 12.2, gate x_5 is a NOR gate with $A = 1$ and $B = 2$ because $w = -3 < 0$, $I_5 = 2 < -w$ and, therefore, $A = -w - I_5 = 3 - 2 = 1$, $B = I_5 = 2$. Since gate x_5 is a NOR gate, from Lemma 12.3, the thresholds of neurons x_3 and x_4 are modified to -7 and 4, respectively. The graph $G_f \setminus \{x_5\}$ is shown in Figure 12.6(a). The edge-weight T_{34} is modified to 0 and therefore, edge (x_3, x_4) does not appear in Figure 12.6(a). With the removal of the edge, x_3 and x_4 become degree two vertices. Since these vertices have equal-weighted edges incident on them, we include them in list L. List L now has two elements, vertices x_3 and x_4.

   Set $G_f \leftarrow G_f \setminus x_5$.

4. Remove vertex x_3 from L, i.e., $L \leftarrow L \setminus x_3$.

5. Delete gate x_3 from G_f. From Lemma 12.2, gate x_3 is an AND gate with $A = 2$ and $B = 3$. From Lemma 12.3, the thresholds of x_1 and x_2 remain unchanged and edge-weight T_{12} is modified to -4. The graph $G_f \setminus \{x_3\}$ is shown in Figure 12.6(b). The removal of vertex x_3 leaves x_1 and x_2 as degree two vertices. However, these vertices are not included in list L since they do not have equal-weighted edges incident on them. List L now has only one vertex x_4.

Set $G_f \leftarrow G_f \setminus x_3$.

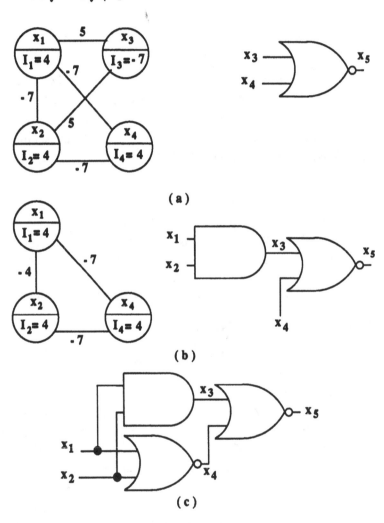

FIGURE 12.6: Transformation of G_f into a combinational circuit.

6. Remove vertex x_4 from L, i.e., $L \leftarrow L \setminus x_4$.

7. Delete gate x_4 from G_f. From Lemma 12.2, gate x_4 is a NOR gate with $A = 3$ and $B = 4$. From Lemma 12.3, the thresholds of x_1 and x_2 are both modified to 0 and the edge-weight T_{12} becomes 0. Therefore, the graph $G_f \setminus \{x_4\}$ consists of two isolated vertices x_1 and x_2, each with a threshold of 0. Since the degree of x_1 and x_2 is zero, we include them in list L which now has two elements, vertices x_1 and x_2.

   Set $G_f \leftarrow G_f \setminus x_4$.

8. Remove vertex x_1 from L, i.e., $L \leftarrow L \setminus x_1$.

9. Label x_1 as a primary input of the combinational circuit since its degree is zero. L now has one element, vertex x_2.

   Set $G_f \leftarrow G_f \setminus x_1$.

10. Remove vertex x_2 from L, i.e., $L \leftarrow L \setminus x_2$.

11. Label x_2 as a primary input of the combinational circuit.

   Set $G_f \leftarrow G_f \setminus x_2$.

12. Stop since L is empty.

The combinational circuit corresponding to G_f is shown in Figure 12.6(c). x_1 and x_2 are the primary inputs of the circuit.

Theorem 12.2: *The minimum of $f(\mathbf{x}) = \mathbf{x}^T Q \mathbf{x} + \mathbf{c}^T \mathbf{x}$ can be found in linear time when the graph G_f is transformable into a combinational circuit of inverters and 2-input AND, OR, NAND and NOR logic gates.*

Proof: From Theorem 12.1, we can determine in $O(n)$ time, where n is the number of variables in f, whether or not f can be transformed into a combinational circuit. Logic simulation of the combinational circuit for an arbitrary set of primary input signal values yields logic values for all signals in the circuit. The value of a variable in f is the logic value of the corresponding signal in the logic circuit. Therefore, the values of all the variables in f are known and an evaluation of f yields the minimum. Since the combinational circuit has n signals, logic simulation can be performed in $O(n)$ time. Therefore, a minimum of pseudo-Boolean quadratic function f can be determined in $O(n)$ time. ∎

Example 12.7: Consider Equation 12.4. The logic circuit corresponding to f (Figure 12.6c) has x_1 and x_2 as the primary inputs. Logic simulation of the circuit with $x_1 = 1$ and $x_2 = 0$ yields $x_3 = 0$, $x_4 = 0$ and $x_5 = 1$. Substitution of these values into f gives 0, the minimum of the pseudo-Boolean quadratic function f.

If the logic circuit corresponding to the function f has p primary inputs, then it has 2^p consistent labelings (Section 12.2). Since only a consistent labeling of the circuit corresponds to a minimum of f, the function has 2^p minimizing points.

The constructive proof of Theorem 12.1 can easily be extended to yield a polynomial time algorithm for transforming G_f into a combinational circuit that contains XOR and XNOR gates as well. Inclusion of XOR and XNOR gates in the combinational circuit will increase the class of quadratic 0-1 programs that can be formulated as logic simulation problems and, hence, solved in linear time.

If only the minimum value of f is desired and not the values of the variables at the minimum of f, logic simulation of the combinational circuit can be avoided.

Theorem 12.3: *Let* $f(\mathbf{x}) = \mathbf{x}^T Q \mathbf{x} + \mathbf{c}^T \mathbf{x} + F$, *where F is the constant term. Also, let C be the combinational circuit corresponding to G_f and let K be the constant term in the energy function E_C of circuit C. The minimum of f is $F - K$.*

Proof: The procedure for transforming G_f into a logic circuit ensures that both f and E_C have the same quadratic and linear terms. Therefore, $f - F = E_C - K$. Since E_C has a minimum value 0 at all consistent labelings of signals in the logic circuit, the minimum of f is $F - K$.   ∎

The constant K can be computed by a slight modification of the procedure for constructing a combinational circuit from G_f (see Section 12.5.1). Initially, set $K = 0$. Whenever a subgraph in Step 2 (in the Proof of Theorem 12.1) represents a gate, add the constant associated with the energy function of the gate (from Table 5.1) to K. When a logic circuit corresponding to G_f is completely constructed, K will be the constant term associated with the energy function of the logic circuit.

Example 12.8: Consider Equation 12.4 in Example 12.4. The constant $F = 6$. Set $K = 0$ and follow the prescribed procedure to convert G_f into a combinational circuit. When the NOR gate with $A = 1$ and $B = 2$ is constructed, increment K by 2 since the constant associated with the energy function of the NOR gate is 2 (Table 5.1). When the AND gate with $A = 2$ and $B = 3$ is constructed, K is unchanged since the constant associated with the energy function for the AND gate is 0. When the NOR gate with $A = 3$ and $B = 4$ is constructed, increment K by 4. After the termination of the procedure, $K = 6$. The minimum of f is $F - K = 0$.

12.6 Minimizing Special Cases

In this section, we investigate the complexity of minimizing special instances of quadratic 0-1 programming problems. As stated earlier, the problem of minimizing arbitrary quadratic 0-1 programming problem is known to be NP-complete. However, there are three special cases when the problem is known to be polynomial-time solvable. Picard and Ratliff [12] have given a polynomial-time algorithm for solving the special case when all the elements of $\mathbf{Q}$ are non-positive (i.e., zero or negative). Barahona [2] has given a polynomial-time algorithm for the case when the graph G_f is series-parallel. Crama et al [6] generalized this result and proposed a polynomial-time algorithm for the case when the graph G_f is of bounded tree-width.

Here, we study the complexity of minimizing special quadratic 0-1 functions in which all the elements of $\mathbf{Q}$ are positive. We show that the problem of minimizing such special functions is NP-complete.

Theorem 12.4: *Finding the minimum of* $f(\mathbf{x}) = \mathbf{x}^T \mathbf{Q} \mathbf{x} + \mathbf{c}^T \mathbf{x}$ *when all elements of* $\mathbf{Q}$ *are positive is NP-complete.*

Proof: Obviously, the problem is in NP. Hence, we need to show that some NP-complete problem is polynomially transformable to the problem of finding the minimum of a pseudo-Boolean quadratic function f with positive coefficients of all the quadratic terms. We shall transform the NP-complete problem of stuck-at fault test generation in combinational circuits [7] into the problem of finding the minimum of f.

The problem of generating a test for a stuck-at fault in an arbitrary combinational circuit can easily be transformed into a problem of generating a test for a fault in a combinational circuit of only NOR gates or NAND gates since all other logic gates can be realized as combinations of NOR or NAND gates. In Chapter 6 we have shown the construction of the neural network for an arbitrary fault in a combinational circuit. This construction can be carried out in $O(n)$ time where n is the number of signals in the combinational circuit. The neural network for a fault in a combinational circuit of only NOR or NAND gates has an energy function with positive coefficients for all quadratic terms. This is because, from Table 5.1, all these gates have quadratic terms with positive coefficients. Therefore, the problem of test generation in a combinational circuit is polynomially transformable into the problem of finding the minimum of the pseudo-Boolean function f with all coefficients of the quadratic terms being positive. ∎

12.7 Summary

We have presented a new class of quadratic 0-1 programming problem cases that can be solved in linear time. When the pseudo-Boolean quadratic function f can be transformed into a combinational logic circuit, then the minimum of f can be obtained by identifying the primary inputs of the circuit and then performing logic simulation for an arbitrary set of 0-1 values for the primary inputs. The transformation of f into a combinational logic circuit requires $O(n)$ time, where n is the number of variables in f. Logic simulation of the resultant logic circuit requires $O(n)$ time. Therefore, the minimum of f can be obtained in $O(n)$ time when f is transformable into a combinational logic circuit. We have also shown that the problem of finding the minimum of f in the restricted case when all elements of Q are positive is NP-complete.

References

[1] A. V. Aho, J. E. Hopcroft, and J. D. Ullman. *The Design and Analysis of Computer Algorithms.* Addison-Wesley Publishing Company, Reading, MA, 1974.

[2] F. Barahona. A Solvable Case of Quadratic 0-1 Programming. *Discrete Applied Mathematics,* 13(1):23–26, January 1986.

[3] M. Boulala and J. P. Uhry. Polytope des Independants d'un Graphe Serie-Parallele. *Discrete Mathematics,* 27(3):225–243, September 1979.

[4] M. W. Carter. The Indefinite Zero One Quadratic Problem. *Discrete Applied Mathematics,* 7(1):23–44, January 1984.

[5] S. T. Chakradhar and M. L. Bushnell. A Solvable Case of Quadratic 0-1 Programming. *Discrete Applied Mathematics. To appear.*

[6] Y. Crama, P. Hansen, and B. Jaumard. The Basic Algorithm for Pseudo-Boolean Programming Revisited. Technical Report RRR # 54-88, Rutgers Center for Operations Research (RUTCOR), Rutgers University, NJ 08903, November 1988.

[7] H. Fujiwara. *Logic Testing and Design for Testability.* MIT Press, Cambridge, Massachusetts, 1985.

[8] M. R. Garey and D. S. Johnson. *Computers and Intractability: A Guide to the Theory of NP-Completeness.* W.H. Freeman & Company, San Francisco, 1979.

[9] P. L. Hammer, P. Hansen, and B. Simeone. Roof Duality, Complementation and Persistence in Quadratic 0-1 Optimization. *Mathematical Programming*, 28(2):121–155, February 1984.

[10] P. L. Hammer and B. Simeone. Quadratic Functions of Binary Variables. Technical Report RRR # 20-87, Rutgers Center for Operations Research (RUTCOR), Rutgers University, NJ 08903, June 1987.

[11] J. J. Hopfield. Artificial Neural Networks. *IEEE Circuits and Devices Magazine*, 4(5):3–10, September 1988.

[12] J. C. Picard and H. D. Ratliff. Minimum Cuts and Related Problems. *Networks*, 5(4):357–370, October 1975.

Chapter 13

SOLVING GRAPH PROBLEMS

"A computer can take more time to find the best solution of a traveling salesman problem than what a traveling salesman may take to complete the worst tour."

In this chapter, we present a VLSI solution of a classical graph problem, the *independent set* problem. This problem occurs in many applications including computer-aided design. Our solution is based on a novel transformation of the graph to a logic circuit. The vertices in the graph are encoded with Boolean variables whose relationships are represented in the logic circuit. The transformation is derived from the energy relation of the neural network model of the logic circuit. Each input vector provides one solution of the independent set problem. The independent set consists of only vertices with true encoding. This new methodology has the potential of solving the problem in real time if programmable logic is used [8].

13.1 Background

The conventional art of electronic computing consists of a mix of algorithms, programming and hardware. For solving very complex problems, however, it is advantageous to transform the problem directly into special-purpose hardware. Examples of such methodology are CAD accelerators, signal processing VLSI chips and artificial neural networks. The last example is considered futuristic by some.

Our new solution to the independent set problem uses a direct mapping of the problem onto a logic circuit. The transformation to logic is obtained through a neural network model whose energy function provides the link between the logic circuit and the graph. Incidentally, our solution method shows a definite and direct relationship between the independent set problem and logic circuits.

Problem statement: Consider a graph $G(\mathcal{V}, \mathcal{E})$ where

$$\mathcal{V} = \{x_1, x_2, \ldots, x_n\}$$

and $\mathcal{E}$ are the vertex set and edge set, respectively. Associate a *weight* $w_i \in \Re$ with every vertex $x_i \in \mathcal{V}$. If $e \in \mathcal{E}$ is an edge with extremities x_i and x_j we write (x_i, x_j) to denote the edge e. An *independent set* in graph $G(\mathcal{V}, \mathcal{E})$ is a subset $\mathcal{V}' \subseteq \mathcal{V}$ such that, for all $x_i, x_j \in \mathcal{V}'$, the edge (x_i, x_j) is *not* in $\mathcal{E}$. The sum of weights associated with vertices in $\mathcal{V}'$ is called the *weight* of the independent set. A *maximum weighted independent set* is an independent set of maximum weight and the problem of finding such a set is known to be NP-hard [10].

Practical applications of this problem appear in computer-aided design, information retrieval, signal transmission analysis, classification theory, economics, scheduling and computer vision [2, 3, 4, 9, 13, 14]. In addition, a method that finds large independent sets is often the essential subroutine in practical *graph coloring* algorithms [5, 6, 10, 11].

13.2 Notation and Terminology

We assume a familiarity with posiforms and pseudo-Boolean functions introduced in Section 9.2. The *conflict graph* C_f of a posiform f has a vertex for every term (except the constant term) and there is an edge between two vertices if and only if the corresponding terms have at least one *conflict variable*, *i.e.*, a variable which is complemented in one term and uncomplemented in the other, or vice versa. Furthermore, every vertex in C_f is associated with a weight equal to the coefficient of the corresponding term. For example, consider the posiform $f = 1 + x_1 + 2\overline{x}_2 + 2\overline{x}_1 x_2$. The corresponding conflict graph C_f is shown in Figure 13.1. A vertex is represented by a circle. The corresponding term is indicated in the upper half. The vertex weight is indicated in the lower half. Since f has three terms $(x_1, 2\overline{x}_2$ and $2\overline{x}_1 x_2)$, disregarding the constant term, the conflict graph has three vertices. Furthermore, there is an edge between terms x_1 and $2\overline{x}_1 x_2$ since variable x_1 is a conflict variable, *i.e.*, x_1 appears complemented in the term $2\overline{x}_1 x_2$ and uncomplemented in the term x_1. Similarly, there is an

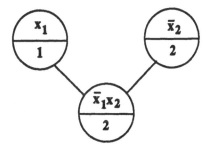

FIGURE 13.1: An example conflict graph.

edge between terms $2\overline{x}_2$ and $2\overline{x}_1 x_2$ since variable x_2 is a conflict variable. For a given graph, a *conflict code* [12] is an assignment of terms to vertices such that adjacent vertices are labeled with terms having at least one conflict variable.

13.3 Maximum Weighted Independent Sets

We will transform the independent set problem into the maximization of a posiform, which in turn, is reduced to the problem of simulating a combinational circuit with an arbitrary primary input vector.

Independent Set to Posiform Maximization: The key idea here is based on the observation that the problem of maximizing a posiform is actually the problem of determining a maximum weighted independent set in the conflict graph of the posiform. The transformation of posiform maximization into an independent set problem is straightforward. Suppose we construct a conflict graph for the posiform. Then, a maximum weighted independent set in the conflict graph can be interpreted as a maximizing point of the posiform.

However, the reverse transformation (independent set to posiform maximization) requires some ingenuity. Given a graph $G(\mathcal{V}, \mathcal{E})$ with certain vertex-weights, we need to label the vertices with terms so that if there is an edge between two vertices, then the corresponding terms have a conflict variable. It can be shown that every graph has a conflict code [12]. As an example, consider the graph shown in Figure 13.2(a). It has three vertices v_1, v_2 and v_3. The vertex weight is indicated in the lower half of the circle representing a vertex. One possible assignment of terms to vertices is shown in Figure 13.2(b). The corresponding posiform is:

$$f = -2 + x_1 + \overline{x}_2 + 2\overline{x}_1 x_2 \qquad (13.1)$$

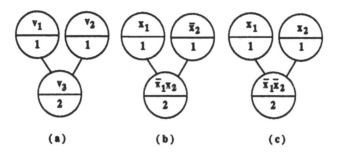

FIGURE 13.2: An example graph (a) and two possible conflict code assignments (b) and (c).

An alternative assignment of terms to vertices is shown in Figure 13.2(c). The corresponding posiform is:

$$f = -3 + x_1 + x_2 + 2\bar{x}_1\bar{x}_2 \qquad (13.2)$$

It can be shown that the global maximizing points of either Equation 13.1 or Equation 13.2 can be interpreted as a maximum weighted independent set of graph $G(\mathcal{V}, \mathcal{E})$.

Posiform maximization to logic simulation: Although every logic simulation problem can be posed as a posiform maximization problem, only a subset of posiform maximization problems can be formulated as instances of the logic simulation problem [7]. A linear time algorithm that determines whether or not a given posiform maximization problem can be formulated as a logic simulation problem is given in Chapter 12. Each variable in the posiform corresponds to a signal in the logic circuit. We simulate the logic circuit with an arbitrary set of 0-1 values at its primary inputs, and the resultant set of signal values corresponds to a global maximizing point of the posiform.

Therefore, the following procedure can be used to isolate a class of independent set problems that are solvable via simulation of a combinational logic circuit:

1. Assign a conflict code to the given graph $G(\mathcal{V}, \mathcal{E})$.

2. Construct a posiform with a conflict graph *isomorphic* [1] to $G(\mathcal{V}, \mathcal{E})$.

3. Transform the posiform into a combinational logic circuit as described in Chapter 12.

4. Simulate the logic circuit with an arbitrary primary input vector to find a set of values for all variables in the posiform. This set of values corresponds to a maximizing point of the posiform.

5. Construct a maximum weighted independent set from the maximizing point of the posiform.

An example in Section 13.5 illustrates this procedure.

13.4 Conflict Graphs of Boolean Gates

For an understanding of the mapping of independent set problem onto a logic circuit, we require some knowledge of the neural network models and conflict graphs of Boolean gates.

Model for a Boolean gate: The energy function for an AND gate with input signals x_1 and x_2, and output signal x_3 is given by (see Chapter 5):

$$
\begin{aligned}
E_{AND}(x_3, x_1, x_2) &= -(A + B)[x_1 x_3 + x_2 x_3] \\
&\quad + B x_1 x_2 + (2A + B) x_3 \qquad (13.3)
\end{aligned}
$$

where $A > 0$ and $B > 0$ are constants. Variables x_1, x_2 and x_3 can assume only binary values. All operations are arithmetic and not Boolean. It is easily verified that only those values of x_1, x_2 and x_3 that are consistent with the function of the AND gate will satisfy $E_{AND} = 0$. Furthermore, $E_{AND} > 0$ for all other combinations of x_1, x_2 and x_3. Note that E_{AND} is a pseudo-Boolean quadratic function that assumes a minimum value of 0 only at values of x_1, x_2 and x_3 consistent with the function of the AND gate. Similar energy functions are derived in Chapter 5 for other logic gates.

We can rewrite Equation 13.3 as follows so that $E_{AND} = 0$ at only those values of x_1, x_2 and x_3 that are consistent with the function of the AND gate and $E_{AND} < 0$ for all other combinations of x_1, x_2 and x_3:

$$
\begin{aligned}
E_{AND}(x_3, x_1, x_2) &= (A + B)[x_1 x_3 + x_2 x_3] \\
&\quad - B x_1 x_2 - (2A + B) x_3 \qquad (13.4)
\end{aligned}
$$

Equation 13.4 can be rewritten as the following posiform:

$$
\begin{aligned}
E_{AND}(x_3, x_1, x_2) &= (A + B)[x_1 x_3 + x_2 x_3] \\
&\quad + B(\overline{x}_1 x_2 + \overline{x}_2) + (2A + B)\overline{x}_3 \\
&\quad - 2(A + B)
\end{aligned}
$$

The corresponding conflict graph is shown in Figure 13.3. Note that Equation 13.4 can also be written as the following posiform:

$$
\begin{aligned}
E_{AND}(x_3, x_1, x_2) &= (A + B)[x_1 x_3 + x_2 x_3] \\
&\quad + B(x_1 \overline{x}_2 + \overline{x}_1) + (2A + B)\overline{x}_3 \\
&\quad - 2(A + B)
\end{aligned}
$$

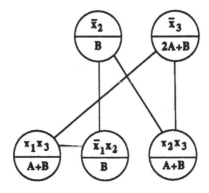

FIGURE 13.3: Conflict graph for an AND gate.

and the corresponding conflict graph is shown in Figure 13.4.

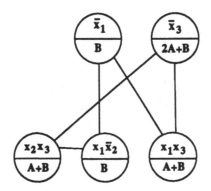

FIGURE 13.4: An alternative conflict graph for the AND gate.

Similarly, the energy function for a NOR gate with inputs x_1 and x_2, and output x_3 can be written as the following posiform:

$$\begin{aligned}
E_{AND}(x_3, x_1, x_2) &= (A + B)[\overline{x}_1 x_3 + \overline{x}_2 x_3] \\
&\quad + B(\overline{x}_1 x_2 + x_1) + (2A + B)\overline{x}_3 \\
&\quad - 2(A + B)
\end{aligned}$$

The corresponding conflict graph is shown in Figure 13.5. Other conflict graphs are also possible for the NOR gate. Note that the conflict graph in Figure 13.5 is isomorphic to the AND gate conflict graphs. Furthermore, the NOR gate conflict graph can be obtained from Figure 13.4 by inverting the literals corresponding to the inputs of the AND gate. Conflict graphs for the OR and NAND gates can be constructed from the conflict graph of an AND gate by appropriately inverting the signals x_1, x_2 and x_3.

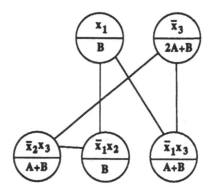

FIGURE 13.5: Conflict graph for the NOR gate.

13.5 An Example

Consider the graph G shown in Figure 13.6. The vertices are labeled v_1, ..., v_{13}. Figure 13.7 shows one possible assignment of terms to vertices.

The posiform corresponding to the conflict graph shown in Figure 13.7 is obtained by the summation of the terms in the graph:

$$
\begin{aligned}
f = \ & 5x_1x_3 + 7\overline{x}_1x_4 + 7\overline{x}_1x_2 + 5x_2x_3 + 7\overline{x}_2x_4 \\
& + 3\overline{x}_3x_5 + 3\overline{x}_4x_5 + 2\overline{x}_3x_4 + 4x_1 + 3\overline{x}_2 \\
& + 5\overline{x}_3 + 10\overline{x}_4 + 4\overline{x}_5 - 28
\end{aligned}
\tag{13.5}
$$

The posiform f can be converted into a combinational circuit by using the linear time algorithm discussed in Section 12.5.1. The corresponding logic circuit is shown in Figure 13.8. The conversion process first eliminates nodes corresponding to the NOR gate x_5 with $A = 1$ and $B = 2$ from the conflict graph in Figure 13.7. The posiform for this NOR gate is as follows:

$$
E_{NOR}(x_5, x_3, x_4) = 3[x_5\overline{x}_3 + x_5\overline{x}_4] + 2[\overline{x}_3x_4 + x_3] + 4\overline{x}_5 - 6
$$

Subtracting the posiform from f yields the following posiform:

$$
\begin{aligned}
f' = \ & 5x_1x_3 + 7\overline{x}_1x_4 + 7\overline{x}_1x_2 + 5x_2x_3 + 7\overline{x}_2x_4 \\
& + 4x_1 + 3\overline{x}_2 + 7\overline{x}_3 + 10\overline{x}_4 - 24
\end{aligned}
$$

The conflict graph of f' is shown in Figure 13.9.

We then eliminate nodes corresponding to the NOR gate x_4 with $A = 3$ and $B = 4$. The posiform for this NOR gate is as follows:

$$
\begin{aligned}
E_{NOR}(x_4, x_1, x_2) = \ & 7[\overline{x}_1x_4 + \overline{x}_2x_4] + 4[\overline{x}_1x_2 + x_1] \\
& + 10\overline{x}_4 - 14
\end{aligned}
$$

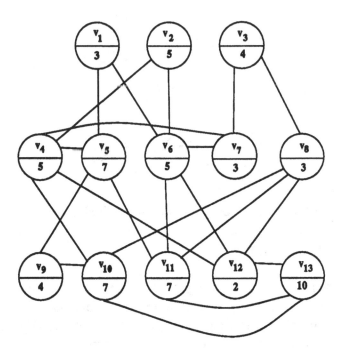

FIGURE 13.6: An example graph $G(\mathcal{V}, \mathcal{E})$.

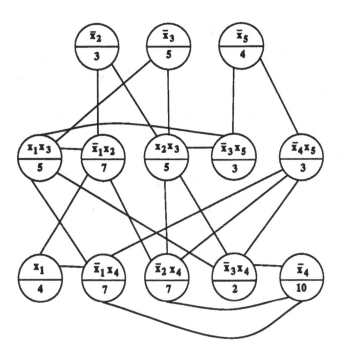

FIGURE 13.7: A possible conflict code for the graph G in Figure 13.6.

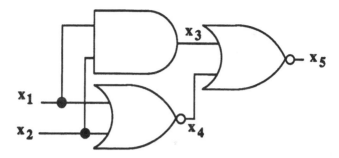

FIGURE 13.8: Logic circuit corresponding to the independent set problem posed on the graph in Figure 13.6.

Subtracting the posiform from f' yields:

$$\begin{aligned} f'' &= 5x_1x_3 + 3\overline{x}_1x_2 + 5x_2x_3 \\ &\quad + 3\overline{x}_2 + 7\overline{x}_3 - 10 \end{aligned}$$

The conflict graph for f'' is shown in Figure 13.10.

Finally, we eliminate nodes corresponding to the AND gate x_3 with $A = 2$ and $B = 3$, from the conflict graph of f''. The posiform of the AND gate is as follows:

$$\begin{aligned} E_{AND}(x_3, x_1, x_2) &= 5x_1x_3 + 3\overline{x}_1x_2 + 5x_2x_3 \\ &\quad + 3\overline{x}_2 + 7\overline{x}_3 - 10 \end{aligned} \qquad (13.6)$$

Subtracting the posiform from f'' yields 0 and the posiform f has been transformed into a logic circuit.

The logic circuit has two primary inputs, x_1 and x_2. Therefore, it can have four input vectors: $(x_1 = x_2 = 0)$, $(x_1 = 0, x_2 = 1)$, $(x_1 = 1, x_2 = 0)$ and $(x_1 = x_2 = 1)$. Each input vector results in a maximizing point of the posiform f. For example, simulation of the circuit with the input vector $(x_1 = x_2 = 0)$ results in $x_3 = 0$, $x_4 = 1$ and $x_5 = 0$. At these values, the posiform f evaluates to 0 and it can easily be verified that 0 is the maximum value of f. Table 13.1 lists all the four maximizing points of f.

A maximum weighted independent set of G (Figure 13.6) can be obtained from a maximizing point of the posiform f, as follows. Consider a maximizing point $(x_1 = x_2 = x_3 = x_5 = 0, x_4 = 1)$ of the posiform f. At this point, only terms corresponding to vertices v_1, v_2, v_3, v_{10}, v_{11} and v_{12}, i.e., terms $\overline{x}_2$, $\overline{x}_3$, $\overline{x}_5$, $\overline{x}_1x_4$, $\overline{x}_2x_4$ and $\overline{x}_3x_4$, respectively, evaluate to a true (1) value. These vertices form the maximum weighted independent set of weight 28. Table 13.1 lists all other maximum weighted independent sets.

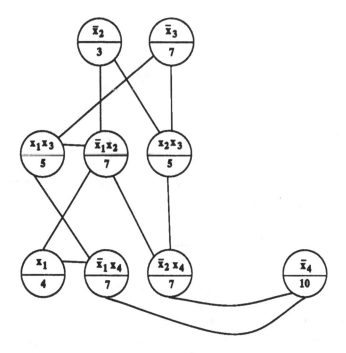

FIGURE 13.9: Conflict graph of posiform f'.

13.6 Summary

We have presented a new methodology for solving problems whose solutions lie in multi-dimensional 0-1 state space. Our transformation to the logic domain gives the potential for implementing such solutions in VLSI. This type of solutions for the traveling salesman problem, circuit placement, logic partitioning, and other problems may also be possible.

As we pointed out, the transformation between the graph and logic is through a posiform expression. This expression has the same form as the energy function of a neural network. In the previous chapters, we pre-

TABLE 13.1: Maximum weighted independent sets of graph G shown in Figure 13.6.

x_1	x_2	x_3	x_4	x_5	f	Independent Sets
0	0	0	1	0	0	$v_1, v_2, v_3, v_{10}, v_{11}, v_{12}$
0	1	0	0	1	0	$v_2, v_5, v_7, v_8, v_{13}$
1	0	0	0	1	0	$v_1, v_2, v_7, v_8, v_9, v_{13}$
1	1	1	0	0	0	$v_3, v_4, v_6, v_9, v_{13}$

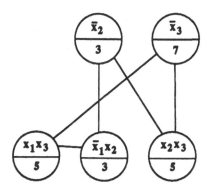

FIGURE 13.10: Conflict graph of posiform f''.

sented energy minimization solutions to the test generation problem. By combining these results, in the future, it may become possible to establish an equivalence between direct-search (e.g., D-algorithm, Podem), energy minimization (including simulated annealing), and logic transformation (as presented in this chapter) methods. An obvious advantage would be that a problem can be transformed to the domain where the most suitable solution method is available.

References

[1] A. V. Aho, J. E. Hopcroft, and J. D. Ullman. *The Design and Analysis of Computer Algorithms*. Addison-Wesley Publishing Company, Reading, MA, 1974.

[2] G. Avondo-Bodeno. *Economic Applications of the Theory of Graphs*. Gordon and Breach Science Publishers, 1962.

[3] E. Balas and C. S. Yu. Finding a Maximum Clique in an Arbitrary Graph. *SIAM Journal on Computing*, 15(6):1054–1068, November 1986.

[4] C. Berge. *The Theory of Graphs and its Applications*. Methuen, 1962.

[5] B. Bollobas and A. Thompson. Random Graphs of Small Order. *Annals of Discrete Mathematics*, 28:47–97, 1985.

[6] R. Boppana and M. Halldórsson. Approximating Maximum Independent Sets by Excluding Subgraphs. In *Proc. of 2nd Scand. Workshop on Algorithm Theory. Springer-Verlag Lecture Notes in Computer Science #447*, pages 13–25, July 1990.

[7] S. T. Chakradhar. *Neural Network Models and Optimization Methods for Digital Testing*. PhD thesis, Department of Computer Science, Rutgers University, New Brunswick, NJ, DCS-TR-269, October 1990.

[8] S. T. Chakradhar and V. D. Agrawal. A Novel VLSI Solution to a Difficult
     Graph Problem. In *Proceedings of the 4th CSI/IEEE International Sympo-
     sium on VLSI Design*, pages 124–129, January 1991.

[9] N. Deo. *Graph Theory with Applications to Engineering and Computer Sci-
     ence*. Prentice-Hall, 1974.

[10] M. R. Garey and D. S. Johnson. *Computers and Intractability: A Guide to
     the Theory of NP-Completeness*. W.H. Freeman & Company, San Francisco,
     1979.

[11] M. Halldórsson. A Still Better Performance Guarantee for Approximate
     Graph Coloring. *Information Processing Letters. To appear*. Also available
     as DIMACS Technical Report 90–44, Rutgers University, 1990.

[12] P. L. Hammer and B. Simeone. Quadratic Functions of Binary Variables.
     Technical Report RRR # 20-87, Rutgers Center for Operations Research
     (RUTCOR), Rutgers University, NJ 08903, June 1987.

[13] C. E. Shannon. The Zero-error Capacity of a Noisy Channel. *I. R. E. Trans-
     actions on Information Theory*, IT-2(3):221, 1956.

[14] J. Turner and W. H. Kautz. A Survey of Progress in Graph Theory in the
     Soviet Union. *SIAM Review*, 12:1–68, 1970.

Chapter 14

OPEN PROBLEMS

"One day Pandora found a sealed jar at the back of a cupboard. It was the jar which Prometheus had asked Epimetheus to keep safely hidden and on no account to open. Though Epimetheus ordered Pandora to leave it alone, she broke the seal – as Zeus intended her to do. Out came a swarm of nasty winged things called Old Age, Sickness, Insanity, Spite, Passion, Vice, Plague, Famine, and so forth. These stung Pandora and Epimetheus most viciously, afterwards going on to attack Prometheus's mortals (who had until then lived happy, decent lives) and spoil everything for them. However, a bright-winged creature called Hope flew out of the jar last of all..."

– R. Graves in *Greek Gods and Heroes*, (1960)

The research reported in the preceeding chapters opens up a new line of thought relating test generation to general optimization problems. We hope the reader will be tempted to find solutions to other currently open problems including the ones listed below:

- *Parallelization of the graph-theoretic test generation technique.* Recall (see Chapter 9) that in this technique, we split the energy function into two sub-functions. A solution to one sub-function, the homogeneous posiform, is obtained very quickly and we check to see if

175

this solution satisfies the other sub-function. One way to parallelize this technique would be to generate several solutions of the homogeneous posiform in parallel and check these, again in parallel, against the other sub-function. This method holds promise since the solution-generating method as well as the solution-checking phase can easily be parallelized.

- *Enhancement of the basic test generation formulation.* We have presented several enhancements (see Chapters 8, 9 and 10) to the basic formulation and we believe more test generation knowledge can be incorporated into the energy function.

- *Simulation of the neural network on parallel or pipelined computers.* Since neurons can be synchronous and use only local information to update their state, the Connection Machine [3] architecture seems to be most suitable for neural network simulations. Similarly, a pipelined machine like MARS [1], which provides fast logic simulation, can be programmed for this application.

- *Identification of a good initial state of the neural network.* We have empirically observed that the center of the hypercube representing the search space is a good initial state. Further research should lead to a better insight into the selection of good starting points for the gradient descent search and the analog neural network.

- *Extension to sequential circuits.* The work reported here deals with combinational circuits. Recently, Fujiwara [2] has extended our neural models to handle three logic values (0, 1 and X). Extending this work to sequential circuits requires neural models of three-state logic. More work is, however, needed to account for the extra degree of freedom (time) in sequential circuits.

- *Development of a good design-for-testability technique.* We have identified a new, easily-testable class of circuits, namely the (k, K)-circuits (see Chapter 11). Design of (k, K)-circuits is an open problem in the area of synthesis for testability.

- *Isolation of new, easily solvable instances of NP-complete problems.* We have isolated special instances that map into combinational circuits. This is a new method of solving such problems. In Chapter 13 we apply it to solve the maximum weighted independent set problem. We believe that using similar techniques, a much larger class of easily-solvable instances can be identified. Some of these instances may map onto sequential circuits.

References

[1] P. Agrawal and W.J. Dally. A Hardware Logic Simulation System. *IEEE Transactions on Computer-Aided Design*, CAD-9(1):19–29, January 1990.

[2] H. Fujiwara. Three-valued Neural Networks for Test Generation. In *Proceedings of the 20th IEEE International Symposium on Fault Tolerant Computing*, pages 64–71, June 1990.

[3] W. D. Hillis. *The Connection Machine*. The MIT Press, Cambridge, Massachusetts, 1985.

Chapter 15

CONCLUSION

"When the conclusions do not contradict any of the given facts, then all neurons can rest."

Our new and unconventional modeling technique for digital circuits has several advantages. First, since the function of the circuit is captured in mathematical expressions, we were able to invent several new techniques to solve problems like test generation. Second, the non-causal form of the model would allow the use of parallel processing for compute-intensive design automation tasks. This parallel processing potential has not been explored.

Based on the logic circuit models, we devised several novel techniques for solving the test generation problem. We described a new, discrete test generation technique using graph-theoretic and mathematical programming techniques. We exploited the special structure of energy functions obtained from test generation problems and developed an efficient technique for testing combinational circuits. Future efforts will implement this technique on multiprocessor systems. We proposed a new class of test generation algorithms that can exploit fine-grain parallel computing and relaxation techniques. Our approach is radically different from the conventional methods of generating tests for circuits from their gate level description. Massive parallelism inherent in neural networks could be exploited to generate tests. Our results on several combinational circuits demonstrated the feasibility of both of the above approaches. With advances in technology,

179

large scale neural networks will become a reality and our neural network approach should provide a significant advantage over other methods.

Since the general problem of stuck-at fault testing is extremely complex, a valid approach is to develop good insights into this problem by considering special classes of easily testable circuits. We have isolated a new class of circuits, called $(k, K)-circuits$, in which the test generation problem is solvable in polynomial time. Synthesis of Boolean functions as easier-to-test $(k, K)-$circuits definitely has the potential for further research.

We presented a novel application of the neural models to solving NP-complete problems. We used quadratic 0-1 programming as a representative NP-complete problem and identified new, linear-time solvable instances of this problem.

Index